Oberwolfach Seminars
Volume 39

Mathias Drton
Bernd Sturmfels
Seth Sullivant

Lectures on Algebraic Statistics

Birkhäuser
Basel · Boston · Berlin

Mathias Drton
University of Chicago
Department of Statistics
5734 S. University Ave
Chicago, IL 60637
USA
e-mail: drton@galton.uchicago.edu

Bernd Sturmfels
Department of Mathematics
University of California
925 Evans Hall
Berkeley, CA 94720
USA
e-mail: bernd@math.berkeley.edu

Seth Sullivant
Department of Mathematics
North Carolina State University
Raleigh, NC 27695-8205
USA
e-mail: seths@math.harvard.edu

2000 Mathematics Subject Classification: 62, 14, 13, 90, 68

Library of Congress Control Number: 2008939526

Bibliographic information published by Die Deutsche Bibliothek
Die Deutsche Bibliothek lists this publication in the Deutsche Nationalbibliografie;
detailed bibliographic data is available in the Internet at <http://dnb.ddb.de>.

ISBN 978-3-7643-8904-8 Birkhäuser Verlag, Basel – Boston – Berlin

© 2009 Birkhäuser Verlag AG
Basel · Boston · Berlin
P.O. Box 133, CH-4010 Basel, Switzerland
Part of Springer Science+Business Media
Printed on acid-free paper produced from chlorine-free pulp. TCF ∞
Printed in Germany

ISBN 978-3-7643-8904-8

e-ISBN 978-3-7643-8905-5

9 8 7 6 5 4 3 2 1

www.birkhauser.ch

Contents

Preface

Algebraic statistics is concerned with the development of techniques in algebraic geometry, commutative algebra, and combinatorics, to address problems in statistics and its applications. On the one hand, algebra provides a powerful tool set for addressing statistical problems. On the other hand, it is rarely the case that algebraic techniques are ready-made to address statistical challenges, and usually new algebraic results need to be developed. This way the dialogue between algebra and statistics benefits both disciplines.

Algebraic statistics is a relatively new field that has developed and changed rather rapidly over the last fifteen years. One of the first pieces of work in this area was the paper of Diaconis and the second author [33], which introduced the notion of a Markov basis for log-linear statistical models and showed its connection to commutative algebra. From there, the algebra/statistics connection spread to a number of different areas including the design of experiments (highlighted in the monograph [74]), graphical models, phylogenetic invariants, parametric inference, algebraic tools for maximum likelihood estimation, and disclosure limitation, to name just a few. References to this literature are surveyed in the editorial [47] and the two review articles [4, 41] in a special issue of the journal *Statistica Sinica*. An area where there has been particularly strong activity is in applications to computational biology, which is highlighted in the book *Algebraic Statistics for Computational Biology* of Lior Pachter and the second author [73]. We will sometimes refer to that book as the "ASCB book."

These lecture notes arose out of a five-day Oberwolfach Seminar, given at the *Mathematisches Forschungsinstitut Oberwolfach* (MFO), in Germany's Black Forest, over the days May 12–16, 2008. The seminar lectures provided an introduction to some of the fundamental notions in algebraic statistics, as well as a snapshot of some of the current research directions. Given such a short timeframe, we were forced to pick and choose topics to present, and many areas of active research in algebraic statistics have been left out. Still, we hope that these notes give an overview of some of the main ideas in the area and directions for future research.

The lecture notes are an expanded version of the thirteen lectures we gave throughout the week, with many more examples and background material than we could fit into our hour-long lectures. The first five chapters cover the material in those thirteen lectures and roughly correspond to the five days of the workshop. Chapter 1 reviews statistical tests for contingency table analysis and explains the notion of a Markov basis for a log-linear model. We connect this notion to commutative algebra, and give some of the most important structural theorems about Markov bases. Chapter 2 is concerned with likelihood inference in algebraic statistical models. We introduce these models for discrete and normal random variables, explain how to solve the likelihood equations parametrically and implicitly, and show how model geometry connects to asymptotics of likelihood ratio statistics. Chapter 3 is an algebraic study of conditional independence structures. We introduce these generally, and then focus in on the special class of graphical models. Chapter 4 is an introduction to hidden variable models. From the algebraic point of view, these models often give rise to secant varieties. Finally, Chapter 5 concerns Bayesian integrals, both from an asymptotic large-sample perspective and from the standpoint of exact evaluation for small samples.

During our week in Oberwolfach, we held several student problem sessions to complement our lectures. We created eight problems highlighting material from the different lectures and assigned the students into groups to work on these problems. The exercises presented a range of computational and theoretical challenges. After daily and sometimes late-night problem solving sessions, the students wrote up solutions, which appear in Chapter 6. On the closing day of the workshop, we held an open problem session, where we and the participants presented open research problems related to algebraic statistics. These appear in Chapter 7.

There are many people to thank for their help in the preparation of this book. First, we would like to thank the MFO and its staff for hosting our Oberwolfach Seminar, which provided a wonderful environment for our research lectures. In particular, we thank MFO director Gert-Martin Greuel for suggesting that we prepare these lecture notes. Second, we thank Birkhäuser editor Thomas Hempfling for his help with our manuscript. Third, we acknowledge support by grants from the U.S. National Science Foundation (Drton DMS-0746265; Sturmfels DMS-0456960; Sullivant DMS-0700078 and 0840795). Bernd Sturmfels was also supported by an Alexander von Humboldt research prize at TU Berlin. Finally, and most importantly, we would like to thank the participants of the seminar. Their great enthusiasm and energy created a very stimulating environment for teaching this material. The participants were Florian Block, Dustin Cartwright, Filip Cools, Jörn Dannemann, Alex Engström, Thomas Friedrich, Hajo Holzmann, Thomas Kahle, Anna Kedzierska, Martina Kubitzke, Krzysztof Latuszynski, Shaowei Lin, Hugo Maruri-Aguilar, Sofia Massa, Helene Neufeld, Mounir Nisse, Johannes Rauh, Christof Söger, Carlos Trenado, Oliver Wienand, Zhiqiang Xu, Or Zuk, and Piotr Zwiernik.

Chapter 1

Markov Bases

This chapter introduces the fundamental notion of a Markov basis, which represents one of the first connections between commutative algebra and statistics. This connection was made in the paper by Diaconis and the second author [33] on contingency table analysis. Statistical hypotheses about contingency tables can be tested in an exact approach by performing random walks on a constrained set of tables with non-negative integer entries. Markov bases are of key importance to this statistical methodology because they comprise moves between tables that ensure that the random walk connects every pair of tables in the considered set.

Section 1.1 reviews the basics of contingency tables and exact tests; for more background see also the books by Agresti [1], Bishop, Holland, Fienberg [18], or Christensen [21]. Section 1.2 discusses Markov bases in the context of hierarchical log-linear models. The problem of computing Markov bases is addressed in Section 1.3, where the problem is placed into the setting of integer lattices and tied to the algebraic notion of a lattice ideal.

1.1 Hypothesis Tests for Contingency Tables

A contingency table contains counts obtained by cross-classifying observed cases according to two or more discrete criteria. Here the word 'discrete' refers to criteria with a finite number of possible levels. As an example consider the 2×2-contingency table shown in Table 1.1.1. This table, which is taken from [1, §5.2.2], presents a classification of 326 homicide indictments in Florida in the 1970s. The two binary classification criteria are the defendant's race and whether or not the defendant received the death penalty. A basic question of interest for this table is whether at the time death penalty decisions were made independently of the defendant's race. In this section we will discuss statistical tests of such independence hypotheses as well as generalizations for larger tables.

	Death Penalty		
Defendant's Race	Yes	No	Total
White	19	141	160
Black	17	149	166
Total	36	290	326

Table 1.1.1: Data on death penalty verdicts.

Classifying a randomly selected case according to two criteria with r and c levels, respectively, yields two random variables X and Y. We code their possible outcomes as $[r]$ and $[c]$, where $[r] := \{1, 2, \ldots, r\}$ and $[c] := \{1, 2, \ldots, c\}$. All probabilistic information about X and Y is contained in the *joint probabilities*

$$p_{ij} = P(X = i, Y = j), \quad i \in [r], \; j \in [c],$$

which determine in particular the *marginal probabilities*

$$p_{i+} := \sum_{j=1}^{c} p_{ij} = P(X = i), \quad i \in [r],$$

$$p_{+j} := \sum_{i=1}^{r} p_{ij} = P(Y = j), \quad j \in [c].$$

Definition 1.1.1. The two random variables X and Y are *independent* if the joint probabilities factor as $p_{ij} = p_{i+}p_{+j}$ for all $i \in [r]$ and $j \in [c]$. We use the symbol $X \perp\!\!\!\perp Y$ to denote independence of X and Y.

Proposition 1.1.2. *The two random variables X and Y are independent if and only if the $r \times c$-matrix $p = (p_{ij})$ has rank 1.*

Proof. ($\Longrightarrow$): The factorization in Definition 1.1.1 writes the matrix p as the product of the column vector filled with the marginal probabilities p_{i+} and the row vector filled with the probabilities p_{+j}. It follows that p has rank 1.

($\Longleftarrow$): Since p has rank 1, it can be written as $p = ab^t$ for $a \in \mathbb{R}^r$ and $b \in \mathbb{R}^c$. All entries in p being non-negative, a and b can be chosen to have non-negative entries as well. Let a_+ and b_+ be the sums of the entries in a and b, respectively. Then, $p_{i+} = a_i b_+$, $p_{+j} = a_+ b_j$, and $a_+ b_+ = 1$. Therefore, $p_{ij} = a_i b_j = a_i b_+ a_+ b_j = p_{i+} p_{+j}$ for all i, j. $\square$

Suppose now that we randomly select n cases that give rise to n independent pairs of discrete random variables

$$\binom{X^{(1)}}{Y^{(1)}}, \binom{X^{(2)}}{Y^{(2)}}, \ldots, \binom{X^{(n)}}{Y^{(n)}} \tag{1.1.1}$$

that are all drawn from the same distribution, that is,

$$P(X^{(k)} = i, Y^{(k)} = j) = p_{ij} \quad \text{for all } i \in [r], \; j \in [c], \; k \in [n].$$

The joint probability matrix $p = (p_{ij})$ for this distribution is considered to be an *unknown* element of the $rc - 1$ dimensional probability simplex

$$\Delta_{rc-1} = \left\{ q \in \mathbb{R}^{r \times c} \; : \; q_{ij} \geq 0 \text{ for all } i, j \text{ and } \sum_{i=1}^{r} \sum_{j=1}^{c} q_{ij} = 1 \right\}.$$

A *statistical model* $\mathcal{M}$ is a subset of Δ_{rc-1}. It represents the set of all candidates for the unknown distribution p.

Definition 1.1.3. The *independence model* for X and Y is the set

$$\mathcal{M}_{X \perp\!\!\!\perp Y} = \{ p \in \Delta_{rc-1} \; : \; \text{rank}(p) = 1 \}.$$

The independence model $\mathcal{M}_{X \perp\!\!\!\perp Y}$ is the intersection of the probability simplex Δ_{rc-1} and the set of all matrices $p = (p_{ij})$ such that

$$p_{ij} p_{kl} - p_{il} p_{kj} = 0 \tag{1.1.2}$$

for all $1 \leq i < k \leq r$ and $1 \leq j < l \leq c$. The solution set to this system of quadratic equations is known as the *Segre variety* in algebraic geometry. If all probabilities are positive, then the vanishing of the 2×2-minor in (1.1.2) corresponds to

$$\frac{p_{ij}/p_{il}}{p_{kj}/p_{kl}} = 1. \tag{1.1.3}$$

Ratios of probabilities being termed *odds*, the ratio in (1.1.3) is known as an *odds ratio* in the statistical literature.

The order of the observed pairs in (1.1.1) carries no information about p and we summarize the observations in a table of counts

$$U_{ij} = \sum_{k=1}^{n} 1_{\{X^{(k)}=i, \, Y^{(k)}=j\}}, \quad i \in [r], \; j \in [c]. \tag{1.1.4}$$

The table $U = (U_{ij})$ is a *two-way contingency table*. We denote the set of all contingency tables that may arise for fixed sample size n by

$$\mathcal{T}(n) := \left\{ u \in \mathbb{N}^{r \times c} \; : \; \sum_{i=1}^{r} \sum_{j=1}^{c} u_{ij} = n \right\}.$$

Proposition 1.1.4. *The random table* $U = (U_{ij})$ *has a* multinomial distribution, *that is, if* $u \in \mathcal{T}(n)$ *and* n *is fixed, then*

$$P(U = u) = \frac{n!}{u_{11}! u_{12}! \cdots u_{rc}!} \prod_{i=1}^{r} \prod_{j=1}^{c} p_{ij}^{u_{ij}}.$$

Proof. We observe $U = u$ if and only if the observations in (1.1.1) include each pair $(i, j) \in [r] \times [c]$ exactly u_{ij} times. The product $\prod_i \prod_j p_{ij}^{u_{ij}}$ is the probability of observing one particular sequence containing each (i, j) exactly u_{ij} times. The pre-multiplied multinomial coefficient is the number of possible sequences of samples that give rise to the counts u_{ij}. $\square$

Consider now the *hypothesis testing* problem

$$H_0 : p \in \mathcal{M}_{X \perp\!\!\!\perp Y} \quad \text{versus} \quad H_1 : p \notin \mathcal{M}_{X \perp\!\!\!\perp Y}. \tag{1.1.5}$$

In other words, we seek to decide whether or not the contingency table U provides evidence against the *null hypothesis H_0*, which postulates that the unknown joint distribution p belongs to the independence model $\mathcal{M}_{X \perp\!\!\!\perp Y}$. This is the question of interest in the death penalty example in Table 1.1.1, and we present two common approaches to this problem.

Chi-square test of independence. If H_0 is true, then $p_{ij} = p_{i+}p_{+j}$, and the expected number of occurrences of the joint event $\{X = i, Y = j\}$ is $np_{i+}p_{+j}$. The two sets of marginal probabilities can be estimated by the corresponding empirical proportions

$$\hat{p}_{i+} = \frac{U_{i+}}{n} \quad \text{and} \quad \hat{p}_{+j} = \frac{U_{+j}}{n},$$

where the *row total*

$$U_{i+} = \sum_{j=1}^{c} U_{ij}$$

counts how often the event $\{X = i\}$ occurred in our data, and the similarly defined *column total* U_{+j} counts the occurrences of $\{Y = j\}$. We can thus estimate the expected counts $np_{i+}p_{+j}$ by $\hat{u}_{ij} = n\hat{p}_{i+}\hat{p}_{+j}$. The *chi-square statistic*

$$X^2(U) = \sum_{i=1}^{r} \sum_{j=1}^{c} \frac{(U_{ij} - \hat{u}_{ij})^2}{\hat{u}_{ij}} \tag{1.1.6}$$

compares the expected counts $\hat{u}_{ij}$ to the observed counts U_{ij} taking into account how likely we estimate each joint event to be. Intuitively, if the null hypothesis is true, we expect X^2 to be small since U should be close to $\hat{u}$. The *chi-square test* rejects the hypothesis H_0, if the statistic X^2 comes out to be "too large."

What is "too large"? This can be gauged using a probability calculation. Let $u \in \mathcal{T}(n)$ be a contingency table containing observed numerical values such as, for instance, Table 1.1.1. Let $X^2(u)$ be the corresponding numerical evaluation of the chi-square statistic. We would like to compute the probability that the random variable $X^2(U)$ defined in (1.1.6) takes a value greater than or equal to $X^2(u)$ provided that H_0 is true. This probability is the *p-value* of the test. If the *p*-value is very small, then it is unlikely to observe a table with chi-square statistic value as large or larger than $X^2(u)$ when drawing data from a distribution in the independence model $\mathcal{M}_{X \perp\!\!\!\perp Y}$. A small p-value thus presents evidence against H_0.

Suppose the p-value for our data is indeed very small, say 0.003. Then, assuming that the model specified by the null hypothesis H_0 is true, the chance of observing data such as those we were presented with or even more extreme is only 3 in 1000. There are now two possible conclusions. Either we conclude that this rare event with probability 0.003 did indeed occur, or we conclude that the null hypothesis was wrong. Which conclusion one is willing to adopt is a subjective decision. However, it has become common practice to reject the null hypothesis if the p-value is smaller than a threshold on the order of 0.01 to 0.05. The latter choice of 0.05 has turned into a default in the scientific literature.

On the other hand, if $X^2(u)$ is deemed to be small, so that the p-value is large, the chi-square test is inconclusive. In this case, we say that the chi-square test does not provide evidence against the null hypothesis.

The above strategy cannot be implemented as such because the probability distribution of $X^2(U)$ depends on where in the model $\mathcal{M}_{X \perp\!\!\!\perp Y}$ the unknown underlying joint distribution $p = (p_{ij})$ lies. However, this problem disappears when considering limiting distributions for growing sample size n.

Definition 1.1.5. The *standard normal distribution* $\mathcal{N}(0,1)$ is the probability distribution on the real line $\mathbb{R}$ that has the density function

$$f(x) = \frac{1}{\sqrt{2\pi}} e^{-x^2/2}.$$

If $Z_1, \ldots, Z_m$ are independent $\mathcal{N}(0,1)$-random variables, then $Z_1^2 + \cdots + Z_m^2$ has a *chi-square distribution* with m *degrees of freedom*, which we denote by χ_m^2.

In the following proposition, we denote the chi-square statistic computed from an n-sample by $X_n^2(U)$ in order to emphasize the dependence on the sample size. A proof of this proposition can be found, for example, in [1, §12.3.3].

Proposition 1.1.6. *If the joint distribution of X and Y is determined by an $r \times c$-matrix $p = (p_{ij})$ in the independence model $\mathcal{M}_{X \perp\!\!\!\perp Y}$ and has positive entries, then*

$$\lim_{n \to \infty} P(X_n^2(U) \geq t) = P(\chi_{(r-1)(c-1)}^2 \geq t) \qquad \text{for all } t > 0.$$

We denote such convergence in distribution by $X_n^2(U) \xrightarrow{D} \chi_{(r-1)(c-1)}^2$.

In this proposition, the shorthand $P(\chi_{(r-1)(c-1)}^2 \geq t)$ denotes the probability $P(W \geq t)$ for a random variable W that follows a chi-square distribution with $(r-1)(c-1)$ degrees of freedom. We will continue to use this notation in subsequent statements about chi-square probabilities.

Each matrix p in the independence model $\mathcal{M}_{X \perp\!\!\!\perp Y}$ corresponds to a pair of two marginal distributions for X and Y, which are in the probability simplices Δ_{r-1} and Δ_{c-1}, respectively. Therefore, the dimension of $\mathcal{M}_{X \perp\!\!\!\perp Y}$ is $(r-1) + (c-1)$. The *codimension* of $\mathcal{M}_{X \perp\!\!\!\perp Y}$ is the difference between the dimensions of the underlying probability simplex Δ_{rc-1} and the model $\mathcal{M}_{X \perp\!\!\!\perp Y}$. We see that the degrees of freedom for the limiting chi-square distribution are given by the codimension $(rc-1) - (r-1) - (c-1) = (r-1)(c-1)$.

The convergence in distribution in Proposition 1.1.6 suggests that we gauge the size of an observed value $X^2(u)$ by computing the probability

$$P(\chi^2_{(r-1)(c-1)} \geq X^2(u)), \tag{1.1.7}$$

which is referred to as the *p-value* for the chi-square test of independence.

Example 1.1.7. For the death penalty example in Table 1.1.1, $r = c = 2$ and the degrees of freedom are $(r-1)(c-1) = 1$. The p-value in (1.1.7) can be computed using the following piece of code for the statistical software R [75]:

```
> u = matrix(c(19,17,141,149),2,2)
> chisq.test(u,correct=FALSE)


        Pearson's Chi-squared test

data:  u
X-squared = 0.2214, df = 1, p-value = 0.638
```

The p-value being large, there is no evidence against the independence model. □

We next present an alternative approach to the testing problem (1.1.5). This approach is *exact* in that it avoids asymptotic considerations.

Fisher's exact test. We now consider 2×2-contingency tables. In this case, the distribution of U loses its dependence on the unknown joint distribution p when we condition on the row and column totals.

Proposition 1.1.8. *Suppose $r = c = 2$. If $p = (p_{ij}) \in \mathcal{M}_{X \perp\!\!\!\perp Y}$ and $u \in \mathcal{T}(n)$, then the conditional distribution of U_{11} given $U_{1+} = u_{1+}$ and $U_{+1} = u_{+1}$ is the hypergeometric distribution $HypGeo(n, u_{1+}, u_{+1})$, that is, the probability*

$$P(U_{11} = u_{11} \mid U_{1+} = u_{1+}, U_{+1} = u_{+1}) = \frac{\binom{u_{1+}}{u_{11}} \binom{n-u_{1+}}{u_{+1}-u_{11}}}{\binom{n}{u_{+1}}}$$

for $u_{11} \in \{\max(0, u_{1+} + u_{+1} - n), \ldots, \min(u_{1+}, u_{+1})\}$ and zero otherwise.

Proof. Fix u_{1+} and u_{+1}. Then, as a function of u_{11}, the conditional probability in question is proportional to the joint probability

$$P(U_{11} = u_{11}, U_{1+} = u_{1+}, U_{+1} = u_{+1}) = P(U_{11} = u_{11}, U_{12} = u_{1+} - u_{11},$$
$$U_{21} = u_{+1} - u_{11}, U_{22} = n - u_{1+} - u_{+1} + u_{11}).$$

By Proposition 1.1.4 and after some simplification, this probability equals

$$\binom{n}{u_{1+}}\binom{u_{1+}}{u_{11}}\binom{n-u_{1+}}{u_{+1}-u_{11}}p_{1+}^{u_{1+}}p_{2+}^{n-u_{1+}}p_{+1}^{u_{+1}}p_{+2}^{n-u_{+1}}.$$

Removing factors that do not depend on u_{11}, we see that this is proportional to

$$\binom{u_{1+}}{u_{11}}\binom{n-u_{1+}}{u_{+1}-u_{11}}.$$

Evaluating the normalizing constant using the binomial identity

$$\sum_{u_{11}}\binom{u_{1+}}{u_{11}}\binom{n-u_{1+}}{u_{+1}-u_{11}}=\binom{n}{u_{+1}}$$

yields the claim. $\qquad\square$

Suppose $u \in \mathcal{T}(n)$ is an observed 2×2-contingency table. Proposition 1.1.8 suggests to base the rejection of H_0 in (1.1.5) on the (conditional) p-value

$$P(X^2(U) \geq X^2(u) \,|\, U_{1+} = u_{1+}, U_{+1} = u_{+1}). \tag{1.1.8}$$

This leads to the test known as Fisher's exact test. The computation of the p-value in (1.1.8) amounts to summing the hypergeometric probabilities

$$\frac{\binom{u_{1+}}{v_{11}}\binom{n-u_{1+}}{u_{+1}-v_{11}}}{\binom{n}{u_{+1}}},$$

over all values $v_{11} \in \{\max(0, u_{1+}+u_{+1}-n), \ldots, \min(u_{1+}, u_{+1})\}$ such that the chi-square statistic for the table with entries v_{11} and $v_{12} = u_{1+} - v_{11}$, $v_{21} = u_{+1} - v_{11}$, $v_{22} = n - u_{1+} - u_{+1} + v_{11}$ is greater than or equal to $X^2(u)$, the chi-square statistic value for the observed table.

Fisher's exact test can be based on criteria other than the chi-square statistic. For instance, one could compare a random table U to the observed table u by calculating which of U_{11} and u_{11} is more likely to occur under the hypergeometric distribution from Proposition 1.1.8. The R command `fisher.test(u)` in fact computes the test in this latter form, which can be shown to have optimality properties that we will not detail here. A discussion of the differences of the two criteria for comparing the random table U with the data u can be found in [28].

As presented above, Fisher's exact test applies only to 2×2-contingency tables but the key idea formalized in Proposition 1.1.8 applies more broadly. This will be the topic of the remainder of this section.

Multi-way tables and log-linear models. Let $X_1, \ldots, X_m$ be discrete random variables with X_l taking values in $[r_l]$. Let $\mathcal{R} = \prod_{l=1}^{m}[r_l]$, and define the *joint probabilities*

$$p_i = p_{i_1 \ldots i_m} = P(X_1 = i_1, \ldots, X_m = i_m), \quad i = (i_1, \ldots, i_m) \in \mathcal{R}.$$

These form a *joint probability table* $p = (p_i \,|\, i \in \mathcal{R})$ that lies in the $\#\mathcal{R} - 1$ dimensional probability simplex $\Delta_{\mathcal{R}-1}$. (Note that, as a shorthand, we will often use $\mathcal{R}$ to represent $\#\mathcal{R}$ in superscripts and subscripts.) The interior of $\Delta_{\mathcal{R}-1}$, denoted by $\text{int}(\Delta_{\mathcal{R}-1})$, consists of all strictly positive probability distributions. The following class of models provides a useful generalization of the independence model from Definition 1.1.3; this is explained in more detail in Example 1.2.1.

Definition 1.1.9. Fix a matrix $A \in \mathbb{Z}^{d \times \mathcal{R}}$ whose columns all sum to the same value. The *log-linear model* associated with A is the set of positive probability tables

$$\mathcal{M}_A = \Big\{ p = (p_i) \in \text{int}(\Delta_{\mathcal{R}-1}) \;:\; \log p \in \text{rowspan}(A) \Big\},$$

where $\text{rowspan}(A) = \text{image}(A^T)$ is the linear space spanned by the rows of A. Here $\log p$ denotes the vector whose i-th coordinate is the logarithm of the positive real number p_i. The term *toric model* was used for $\mathcal{M}_A$ in the ASCB book [73, §1.2].

Consider again a set of counts

$$U_i \;=\; \sum_{k=1}^{n} 1_{\{X_1^{(k)} = i_1, \dots, X_m^{(k)} = i_m\}}, \qquad i = (i_1, \dots, i_m) \in \mathcal{R}, \qquad (1.1.9)$$

based on a random n-sample of independent and identically distributed vectors

$$\begin{pmatrix} X_1^{(1)} \\ \vdots \\ X_m^{(1)} \end{pmatrix}, \begin{pmatrix} X_1^{(2)} \\ \vdots \\ X_m^{(2)} \end{pmatrix}, \dots, \begin{pmatrix} X_1^{(n)} \\ \vdots \\ X_m^{(n)} \end{pmatrix}.$$

The counts U_i now form an m-way table $U = (U_i)$ in $\mathbb{N}^{\mathcal{R}}$. Let

$$\mathcal{T}(n) = \Big\{ u \in \mathbb{N}^{\mathcal{R}} \;:\; \sum_{i \in \mathcal{R}} u_i = n \Big\}.$$

Definition 1.1.10. We call the vector Au the *minimal sufficient statistics* for the model $\mathcal{M}_A$, and the set of tables

$$\mathcal{F}(u) = \big\{ v \in \mathbb{N}^{\mathcal{R}} \;:\; Av = Au \big\}$$

is called the *fiber* of a contingency table $u \in \mathcal{T}(n)$ with respect to the model $\mathcal{M}_A$.

Our definition of minimal sufficient statistics is pragmatic. In fact, sufficiency and minimal sufficiency are general statistical notions. When these are applied to the log-linear model $\mathcal{M}_A$, however, one finds that the vector Au is indeed a minimal sufficient statistic in the general sense.

Note that since the row span of A is assumed to contain the vector of 1s, the tables in the fiber $\mathcal{F}(u)$ sum to n. The next proposition highlights the special role played by the sufficient statistics and provides a generalization of Proposition 1.1.8, which drove Fisher's exact test.

Proposition 1.1.11. *If $p = e^{A^T \alpha} \in \mathcal{M}_A$ and $u \in \mathcal{T}(n)$, then*

$$P(U = u) = \frac{n!}{\prod_{i \in \mathcal{R}} u_i!} e^{\alpha^T(Au)},$$

and the conditional probability $P(U = u \mid AU = Au)$ does not depend on p.

Proof. As a generalization of Proposition 1.1.4, it holds that

$$P(U = u) = \frac{n!}{\prod_{i \in \mathcal{R}} u_i!} \prod_{i \in \mathcal{R}} p_i^{u_i} = \frac{n!}{\prod_{i \in \mathcal{R}} u_i!} \prod_{i \in \mathcal{R}} e^{(A^T \alpha)_i u_i} = \frac{n!}{\prod_{i \in \mathcal{R}} u_i!} e^{\alpha^T(Au)}.$$

Moreover,

$$P(U = u \mid AU = Au) = \frac{P(U = u)}{P(AU = Au)},$$

where

$$P(AU = Au) = \sum_{v \in \mathcal{F}(u)} \frac{n!}{\prod_{i \in \mathcal{R}} v_i!} e^{\alpha^T(Av)} = n! \cdot e^{\alpha^T(Au)} \sum_{v \in \mathcal{F}(u)} \left(\prod_{i \in \mathcal{R}} v_i! \right)^{-1}.$$

It follows that

$$P(U = u \mid AU = Au) = \frac{1/\left(\prod_{i \in \mathcal{R}} u_i! \right)}{\sum_{v \in \mathcal{F}(u)} 1/\left(\prod_{i \in \mathcal{R}} v_i! \right)}. \tag{1.1.10}$$

This expression is independent of α and hence independent of p. $\square$

Consider the hypothesis testing problem

$$H_0 : p \in \mathcal{M}_A \quad \text{versus} \quad H_1 : p \notin \mathcal{M}_A. \tag{1.1.11}$$

Based on Proposition 1.1.11, we can generalize Fisher's exact test by computing the p-value

$$P(X^2(U) \geq X^2(u) \mid AU = Au). \tag{1.1.12}$$

Here

$$X^2(U) = \sum_{i \in \mathcal{R}} \frac{(U_i - \hat{u}_i)^2}{\hat{u}_i} \tag{1.1.13}$$

is the natural generalization of the *chi-square statistic* in (1.1.6). Evaluation of $X^2(U)$ requires computing the model-based expected counts $\hat{u}_i = n\hat{p}_i$, where $\hat{p}_i$ are the *maximum likelihood estimates* discussed in Section 2.1. There, it will also become clear that the estimates $\hat{p}_i$ are identical for all tables in a fiber $\mathcal{F}(u)$.

Exact computation of the p-value in (1.1.12) involves summing over all non-negative integer solutions to the system of linear equations in (1.1.10). Indeed, the p-value is equal to

$$\frac{\sum_{v\in\mathcal{F}(u)} 1_{\{X^2(v)\geq X^2(u)\}}/\left(\prod_{i\in\mathcal{R}} u_i!\right)}{\sum_{v\in\mathcal{F}(u)} 1/\left(\prod_{i\in\mathcal{R}} v_i!\right)}.$$

In even moderately sized contingency tables, the exact evaluation of that sum can become prohibitive. However, the p-value can still be approximated using *Markov chain Monte Carlo* algorithms for sampling tables from the conditional distribution of U given $AU = Au$.

Definition 1.1.12. Let $\mathcal{M}_A$ be the log-linear model associated with a matrix A whose integer kernel we denote by $\ker_{\mathbb{Z}}(A)$. A finite subset $\mathcal{B} \subset \ker_{\mathbb{Z}}(A)$ is a *Markov basis* for $\mathcal{M}_A$ if for all $u \in \mathcal{T}(n)$ and all pairs $v, v' \in \mathcal{F}(u)$ there exists a sequence $u_1, \ldots, u_L \in \mathcal{B}$ such that

$$v' = v + \sum_{k=1}^{L} u_k \quad \text{and} \quad v + \sum_{k=1}^{l} u_k \geq 0 \quad \text{for all } l = 1, \ldots, L.$$

The elements of the Markov basis are called *moves*.

The existence and computation of Markov bases will be the subject of Sections 1.2 and 1.3. Once we have found such a Markov basis $\mathcal{B}$ for the model $\mathcal{M}_A$, we can run the following algorithm that performs a random walk on a fiber $\mathcal{F}(u)$.

Algorithm 1.1.13 (Metropolis-Hastings).
Input: *A contingency table $u \in \mathcal{T}(n)$ and a Markov basis $\mathcal{B}$ for the model $\mathcal{M}_A$.*
Output: *A sequence of chi-square statistic values $(X^2(v_t))_{t=1}^{\infty}$ for tables v_t in the fiber $\mathcal{F}(u)$.*
Step 1: *Initialize $v_1 = u$.*
Step 2: *For $t = 1, 2, \ldots$ repeat the following steps:*

(i) *Select uniformly at random a move $u_t \in \mathcal{B}$.*

(ii) *If $\min(v_t + u_t) < 0$, then set $v_{t+1} = v_t$, else set*

$$v_{t+1} = \begin{cases} v_t + u_t \\ v_t \end{cases} \quad \text{with probability} \quad \begin{cases} q \\ 1 - q \end{cases},$$

where

$$q = \min\left\{1, \frac{P(U = v_t + u_t \mid AU = Au)}{P(U = v_t \mid AU = Au)}\right\}.$$

(iii) *Compute $X^2(v_t)$.*

An important feature of the Metropolis-Hasting algorithm is that the probability q in Step 2(ii) is defined as a ratio of two conditional probabilities. Therefore, we never need to evaluate the sum in the denominator in (1.1.10).

Theorem 1.1.14. *The output* $(X^2(v_t))_{t=1}^\infty$ *of Algorithm 1.1.13 is an aperiodic, reversible and irreducible Markov chain that has stationary distribution equal to the conditional distribution of* $X^2(U)$ *given* $AU = Au$.

A proof of this theorem can be found, for example, in [33, Lemma 2.1] or [78, Chapter 6]. It is clear that selecting the proposed moves u_t from a Markov basis ensures the irreducibility (or connectedness) of the Markov chain. The following corollary clarifies in which sense Algorithm 1.1.13 computes the p-value in (1.1.12).

Corollary 1.1.15. *With probability* 1, *the output sequence* $(X^2(v_t))_{t=1}^\infty$ *of Algorithm 1.1.13 satisfies*

$$\lim_{M \to \infty} \frac{1}{M} \sum_{t=1}^M 1_{\{X^2(v_t) \geq X^2(u)\}} = P(X^2(U) \geq X^2(u) \mid AU = Au).$$

A proof of this law of large numbers can be found in [78, Chapter 6], where heuristic guidelines for deciding how long to run Algorithm 1.1.13 are also given; compare [78, Chapter 8]. Algorithm 1.1.13 is only the most basic scheme for sampling tables from a fiber. Instead one could also apply a feasible multiple of a selected Markov basis move. As discussed in [33], this will generally lead to a better mixing behavior of the constructed Markov chain. However, few theoretical results are known about the mixing times of these algorithms in the case of hypergeometric distributions on fibers of contingency tables considered here.

1.2 Markov Bases of Hierarchical Models

Continuing our discussion in Section 1.1, with each matrix $A \in \mathbb{Z}^{d \times \mathcal{R}}$ we associate a log-linear model $\mathcal{M}_A$. This is the set of probability distributions

$$\mathcal{M}_A = \{p \in \text{int}(\Delta_{\mathcal{R}-1}) : \log p \in \text{rowspan}(A)\}.$$

We assume throughout that the sum of the entries in each column of the matrix A is a fixed value.

This section introduces the class of hierarchical log-linear models and describes known results about their Markov bases. Recall that a Markov basis is a special spanning set of the lattice $\ker_{\mathbb{Z}} A$, the integral kernel of A. The Markov basis can be used to perform irreducible random walks over the fibers $\mathcal{F}(u)$.

By a *lattice* we mean a subgroup of the additive group $\mathbb{Z}^{\mathcal{R}}$. Markov bases, and other types of bases, for general lattices will be discussed in Section 1.3. Often we will interchangeably speak of the Markov basis for $\mathcal{M}_A$, the Markov basis for the matrix A, or the Markov basis for the lattice $\ker_{\mathbb{Z}} A := \ker A \cap \mathbb{Z}^{\mathcal{R}}$. These three expressions mean the same thing, and the particular usage depends on the context. Before describing these objects for general hierarchical models, we will first focus on the motivating example from the previous section, namely, the model of independence. This is a special instance of a hierarchical model.

Example 1.2.1 (Independence). An $r \times c$ probability table $p = (p_{ij})$ is in the independence model $\mathcal{M}_{X \perp\!\!\!\perp Y}$ if and only if each p_{ij} factors into the product of the marginal probabilities p_{i+} and p_{+j}. If p has all positive entries, then

$$\log p_{ij} = \log p_{i+} + \log p_{+j}, \quad i \in [r], \ j \in [c]. \tag{1.2.1}$$

For a concrete example, suppose that $r = 2$ and $c = 3$. Then $\log p$ is a 2×3 matrix, but we write this matrix as a vector with six coordinates. Then (1.2.1) states that the vector $\log p$ lies in the row span of the matrix

$$A = \begin{array}{c} \begin{array}{cccccc} 11 & 12 & 13 & 21 & 22 & 23 \end{array} \\ \begin{pmatrix} 1 & 1 & 1 & 0 & 0 & 0 \\ 0 & 0 & 0 & 1 & 1 & 1 \\ 1 & 0 & 0 & 1 & 0 & 0 \\ 0 & 1 & 0 & 0 & 1 & 0 \\ 0 & 0 & 1 & 0 & 0 & 1 \end{pmatrix} \end{array}.$$

We see that the positive part of the independence model is equal to the log-linear model $\mathcal{M}_A$. For general table dimensions, A is an $(r + c) \times rc$ matrix.

Let u be an $r \times c$ table, which we again think of in "vectorized" format. The matrix A that represents the model of independence is determined by the identity

$$Au = \begin{pmatrix} u_{\cdot+} \\ u_{+\cdot} \end{pmatrix},$$

where $u_{\cdot+}$ and $u_{+\cdot}$ are the vectors of row and columns sums of the table u. In the particular instance of $r = 2$ and $c = 3$, the above identity reads

$$Au = \begin{pmatrix} 1 & 1 & 1 & 0 & 0 & 0 \\ 0 & 0 & 0 & 1 & 1 & 1 \\ 1 & 0 & 0 & 1 & 0 & 0 \\ 0 & 1 & 0 & 0 & 1 & 0 \\ 0 & 0 & 1 & 0 & 0 & 1 \end{pmatrix} \begin{pmatrix} u_{11} \\ u_{12} \\ u_{13} \\ u_{21} \\ u_{22} \\ u_{23} \end{pmatrix} = \begin{pmatrix} u_{1+} \\ u_{2+} \\ u_{+1} \\ u_{+2} \\ u_{+3} \end{pmatrix}.$$

The moves to perform the random walk in Fisher's exact test of independence are drawn from the lattice

$$\ker_{\mathbb{Z}} A = \left\{ v \in \mathbb{Z}^{r \times c} : \sum_{k=1}^{r} v_{kj} = 0 \text{ for all } j, \text{ and } \sum_{k=1}^{c} v_{ik} = 0 \text{ for all } i \right\},$$

which consists of all $r \times c$ integer tables whose row and column sums are zero. $\square$

For the standard model of independence of two discrete random variables, the lattice $\ker_{\mathbb{Z}} A$ contains a collection of obvious small vectors. In the Markov basis literature, these moves are often known as *basic moves*. Let e_{ij} denote the standard unit table, which has a 1 in the (i, j) position, and zeroes elsewhere. If u is a vector or matrix, then $\|u\|_1 = \sum_{i=1}^{R} |u_i|$ denotes the 1-norm of u.

Proposition 1.2.2. *The unique minimal Markov basis for the independence model* $\mathcal{M}_{X \perp\!\!\!\perp Y}$ *consists of the following* $2 \cdot \binom{r}{2}\binom{c}{2}$ *moves, each having 1-norm 4:*

$$\mathcal{B} = \{\pm(e_{ij} + e_{kl} - e_{il} - e_{kj}) : 1 \le i < k \le r, 1 \le j < l \le c\}.$$

Proof. Let $u \neq v$ be two non-negative integral tables that have the same row and column sums. It suffices to show that there is an element $b \in \mathcal{B}$, such that $u+b \ge 0$ and $\|u - v\|_1 > \|u + b - v\|_1$, because this implies that we can use elements of $\mathcal{B}$ to bring points in the same fiber closer to one another. Since u and v are not equal and $Au = Av$, there is at least one positive entry in $u - v$. Without loss of generality, we may suppose $u_{11} - v_{11} > 0$. Since $u - v \in \ker_{\mathbb{Z}} A$, there is an entry in the first row of $u - v$ that is negative, say $u_{12} - v_{12} < 0$. By a similar argument $u_{22} - v_{22} > 0$. But this implies that we can take $b = e_{12} + e_{21} - e_{11} - e_{22}$ which attains $\|u - v\|_1 > \|u + b - v\|_1$ and $u + b \ge 0$ as desired.

The Markov basis $\mathcal{B}$ is minimal because, if one of the elements of $\mathcal{B}$ is omitted, the fiber which contains its positive and negative parts will be disconnected. That this minimal Markov basis is unique is a consequence of the characterization of (non)uniqueness of Markov bases in Theorem 1.3.2 below. $\qquad\square$

As preparation for more complex log-linear models, we mention that it is often useful to use a unary representation for the Markov basis elements. That is, we can write a Markov basis element by recording, with multiplicities, the indices of the non-zero entries that appear. This notation is called *tableau notation*.

Example 1.2.3. The tableau notation for the moves in the Markov basis of the independence model is

$$\begin{bmatrix} i & j \\ k & l \end{bmatrix} - \begin{bmatrix} i & l \\ k & j \end{bmatrix},$$

which corresponds to exchanging $e_{ij} + e_{kl}$ with $e_{il} + e_{kj}$. For the move $e_{11} + e_{12} - 2e_{13} - e_{21} - e_{22} + 2e_{23}$, which arises in Exercise 6.1, the tableau notation is

$$\begin{bmatrix} 1 & 1 \\ 1 & 2 \\ 2 & 3 \\ 2 & 3 \end{bmatrix} - \begin{bmatrix} 1 & 3 \\ 1 & 3 \\ 2 & 1 \\ 2 & 2 \end{bmatrix}.$$

Note that the indices 13 and 23 are both repeated twice, since e_{13} and e_{23} both appear with multiplicity 2 in the move. $\qquad\square$

Among the most important classes of log-linear models are the hierarchical log-linear models. In these models, interactions between random variables are encoded by a simplicial complex, whose vertices correspond to the random variables, and whose faces correspond to interaction factors that are also known as *potential functions*. The independence model, discussed above, is the most basic instance of a hierarchical model. We denote the power set of $[m]$ by $2^{[m]}$.

Definition 1.2.4. A *simplicial complex* is a set $\Gamma \subseteq 2^{[m]}$ such that $F \in \Gamma$ and $S \subset F$ implies that $S \in \Gamma$. The elements of Γ are called *faces* of Γ and the inclusion-maximal faces are the *facets* of Γ.

To describe a simplicial complex we need only list its facets. We will use the bracket notation from the theory of hierarchical log-linear models [21]. For instance $\Gamma = [12][13][23]$ is the bracket notation for the simplicial complex

$$\Gamma = \{\emptyset, \{1\}, \{2\}, \{3\}, \{1,2\}, \{1,3\}, \{2,3\}\}.$$

As described above, a log-linear model is defined by a non-negative integer matrix A, and the model $\mathcal{M}_A$ consists of all probability distributions whose co-ordinatewise logarithm lies in the row span of A. If $\log p \in \mathrm{rowspan}(A)$, there is an $\alpha \in \mathbb{R}^d$ such that $\log p = A^T \alpha$. Exponentiating, we have $p = \exp(A^T \alpha)$. It is natural to use this expression as a parametrization for the set of all probability distributions lying in the model, in which case we must introduce a normalizing constant $Z(\alpha)$ to guarantee that we get a probability distribution:

$$p \;=\; \frac{1}{Z(\alpha)} \exp(A^T \alpha).$$

We can make things simpler and more algebraic by avoiding the exponential notation. Instead, we will often use the equivalent *monomial notation* when writing the parametrization of a log-linear model. Indeed, setting $\theta_i = \exp(\alpha_i)$, we have

$$p_j \;=\; P(X = j) \;=\; \frac{1}{Z(\theta)} \cdot \prod_{i=1}^{d} \theta_i^{a_{ij}} \tag{1.2.2}$$

where $A = (a_{ij})$. This monomial expression can be further abbreviated as $\theta^{a_j} = \prod_{i=1}^{d} \theta_i^{a_{ij}}$ where a_j denotes the jth column of A.

The definition of log-linear models depends on first specifying a matrix $A = (a_{ij})$, and then describing a family of probability distributions via the parametrization (1.2.2). For many log-linear models, however, it is easiest to give the monomial parametrization first, and then recover the matrix A and the sufficient statistics. In particular, this is true for the family of hierarchical log-linear models.

We use the following convention for writing subindices. If $i = (i_1, \ldots, i_m) \in \mathcal{R}$ and $F = \{f_1, f_2, \ldots\} \subseteq [m]$ then $i_F = (i_{f_1}, i_{f_2}, \ldots)$. For each subset $F \subseteq [m]$, the random vector $X_F = (X_f)_{f \in F}$ has the state space $\mathcal{R}_F = \prod_{f \in F}[r_f]$.

Definition 1.2.5. Let $\Gamma \subseteq 2^{[m]}$ be a simplicial complex and let $r_1, \ldots, r_m \in \mathbb{N}$. For each facet $F \in \Gamma$, we introduce a set of $\#\mathcal{R}_F$ positive parameters $\theta_{i_F}^{(F)}$. The *hierarchical log-linear model* associated with Γ is the set of all probability distributions

$$\mathcal{M}_\Gamma = \left\{ p \in \Delta_{\mathcal{R}-1} \;:\; p_i = \frac{1}{Z(\theta)} \prod_{F \in \mathrm{facet}(\Gamma)} \theta_{i_F}^{(F)} \text{ for all } i \in \mathcal{R} \right\}, \tag{1.2.3}$$

where $Z(\theta)$ is the normalizing constant (or partition function)

$$Z(\theta) = \sum_{i \in \mathcal{R}} \prod_{F \in \text{facet}(\Gamma)} \theta_{i_F}^{(F)}.$$

Example 1.2.6 (Independence). Let $\Gamma = [1][2]$. Then the hierarchical model consists of all positive probability matrices $(p_{i_1 i_2})$

$$p_{i_1 i_2} = \frac{1}{Z(\theta)} \theta_{i_1}^{(1)} \theta_{i_2}^{(2)}$$

where $\theta^{(j)} \in (0, \infty)^{r_j}$, $j = 1, 2$. That is, the model consists of all positive rank 1 matrices. It is the positive part of the model of independence $\mathcal{M}_{X \perp\!\!\!\perp Y}$, or in algebraic geometric language, the positive part of the Segre variety. □

Example 1.2.7 (No 3-way interaction). Let $\Gamma = [12][13][23]$ be the boundary of a triangle. The hierarchical model $\mathcal{M}_\Gamma$ consists of all $r_1 \times r_2 \times r_3$ tables $(p_{i_1 i_2 i_3})$ with

$$p_{i_1 i_2 i_3} = \frac{1}{Z(\theta)} \theta_{i_1 i_2}^{(12)} \theta_{i_1 i_3}^{(13)} \theta_{i_2 i_3}^{(23)}$$

for some positive real tables $\theta^{(12)} \in (0, \infty)^{r_1 \times r_2}, \theta^{(13)} \in (0, \infty)^{r_1 \times r_3}$, and $\theta^{(23)} \in (0, \infty)^{r_2 \times r_3}$. Unlike the case of the model of independence, this important statistical model does not have a correspondence with any classically studied algebraic variety. In the case of binary random variables, its implicit representation is the equation

$$p_{111} p_{122} p_{212} p_{221} = p_{112} p_{121} p_{211} p_{222}.$$

That is, the log-linear model consists of all positive probability distributions that satisfy this quartic equation. Implicit representations for log-linear models will be explained in detail in Section 1.3, and a general discussion of implicit representations will appear in Section 2.2. □

Example 1.2.8 (Something more general). Let $\Gamma = [12][23][345]$. The hierarchical model $\mathcal{M}_\Gamma$ consists of all $r_1 \times r_2 \times r_3 \times r_4 \times r_5$ probability tensors $(p_{i_1 i_2 i_3 i_4 i_5})$ with

$$p_{i_1 i_2 i_3 i_4 i_5} = \frac{1}{Z(\theta)} \theta_{i_1 i_2}^{(12)} \theta_{i_2 i_3}^{(23)} \theta_{i_3 i_4 i_5}^{(345)},$$

for some positive real tables $\theta^{(12)} \in (0, \infty)^{r_1 \times r_2}, \theta^{(23)} \in (0, \infty)^{r_2 \times r_3}$, and $\theta^{(345)} \in (0, \infty)^{r_3 \times r_4 \times r_5}$. These tables of parameters represent the potential functions. □

To begin to understand the Markov bases of hierarchical models, we must come to terms with the 0/1 matrices A_Γ that realize these models in the form $\mathcal{M}_{A_\Gamma}$. In particular, we must determine what linear transformation the matrix A_Γ represents. Let $u \in \mathbb{N}^{\mathcal{R}}$ be an $r_1 \times \cdots \times r_m$ contingency table. For any subset $F = \{f_1, f_2, \ldots\} \subseteq [m]$, let $u|_F$ be the $r_{f_1} \times r_{f_2} \times \cdots$ marginal table such that $(u|_F)_{i_F} = \sum_{j \in \mathcal{R}_{[m] \backslash F}} u_{i_F, j}$. The table $u|_F$ is called the *F-marginal* of u.

Proposition 1.2.9. *Let* $\Gamma = [F_1][F_2]\cdots$. *The matrix* A_Γ *represents the linear map*

$$u \mapsto (u|_{F_1}, u|_{F_2}, \ldots),$$

and the Γ-marginals are minimal sufficient statistics of the hierarchical model $\mathcal{M}_\Gamma$.

Proof. We can read the matrix A_Γ off the parametrization. In the parametrization, the rows of A_Γ correspond to parameters, and the columns correspond to states. The rows come in blocks that correspond to the facets F of Γ. Each block has cardinality $\#\mathcal{R}_F$. Hence, the rows of A_Γ are indexed by pairs (F, i_F) where F is a facet of Γ and $i_F \in \mathcal{R}_F$. The columns of A_Γ are indexed by all elements of $\mathcal{R}$. The entry in A_Γ for row index (F, i_F) and column index $j \in \mathcal{R}$ equals 1 if $j_F = i_F$ and equals zero otherwise. This description follows by reading the parametrization from (1.2.3) down the column of A_Γ that corresponds to p_j. The description of minimal sufficient statistics as marginals comes from reading this description across the rows of A_Γ, where the block corresponding to F, yields the F-marginal $u|_F$. See Definition 1.1.10. $\square$

Example 1.2.10. Returning to our examples above, for $\Gamma = [1][2]$ corresponding to the model of independence, the minimal sufficient statistics are the row and column sums of $u \in \mathbb{N}^{r_1 \times r_2}$. Thus we have $A_{[1][2]}u = (u|_1, u|_2)$. Above, we abbreviated these row and column sums by $u_{.+}$ and $u_{+.}$, respectively.

For the model of no 3-way interaction, with $\Gamma = [12][13][23]$, the minimal sufficient statistics consist of all 2-way margins of the 3-way table u. That is

$$A_{[12][13][23]}u = (u|_{12}, u|_{13}, u|_{23})$$

and $A_{[12][13][23]}$ is a matrix with $r_1 r_2 + r_1 r_3 + r_2 r_3$ rows and $r_1 r_2 r_3$ columns. $\square$

As far as explicitly writing down the matrix A_Γ, this can be accomplished in a uniform way by assuming that the rows and columns are ordered lexicographically.

Example 1.2.11. Let $\Gamma = [12][14][23]$ and $r_1 = r_2 = r_3 = r_4 = 2$. Then A_Γ equals

1111	1112	1121	1122	1211	1212	1221	1222	2111	2112	2121	2122	2211	2212	2221	2222
1	1	1	1	0	0	0	0	0	0	0	0	0	0	0	0
0	0	0	0	1	1	1	1	0	0	0	0	0	0	0	0
0	0	0	0	0	0	0	0	1	1	1	1	0	0	0	0
0	0	0	0	0	0	0	0	0	0	0	0	1	1	1	1
1	0	1	0	1	0	1	0	0	0	0	0	0	0	0	0
0	1	0	1	0	1	0	1	0	0	0	0	0	0	0	0
0	0	0	0	0	0	0	0	1	0	1	0	1	0	1	0
0	0	0	0	0	0	0	0	0	1	0	1	0	1	0	1
1	1	0	0	0	0	0	0	1	1	0	0	0	0	0	0
0	0	1	1	0	0	0	0	0	0	1	1	0	0	0	0
0	0	0	0	1	1	0	0	0	0	0	0	1	1	0	0
0	0	0	0	0	0	1	1	0	0	0	0	0	0	1	1

where the rows correspond to ordering the facets of Γ in the order listed above and using the lexicographic ordering $11 > 12 > 21 > 22$ within each facet. $\qquad\square$

Now that we know how to produce the matrices A_Γ, we can begin to compute examples of Markov bases. The program 4ti2 [57] computes a Markov basis of a lattice $\ker_{\mathbb{Z}}(A)$ taking as input either the matrix A or a spanning set for $\ker_{\mathbb{Z}} A$. By entering a spanning set as input, 4ti2 can also be used to compute Markov bases for general lattices $\mathcal{L}$ (see Section 1.3). A repository of Markov bases for a range of widely used hierarchical models is being maintained by Thomas Kahle and Johannes Rauh at http://mbdb.mis.mpg.de/.

Example 1.2.12. We use 4ti2 to compute the Markov basis of the no 3-way interaction model $\Gamma = [12][13][23]$, for three binary random variables $r_1 = r_2 = r_3 = 2$. The matrix representing this model has format 12×8. First, we create a file no3way which is the input file consisting of the size of the matrix, and the matrix itself:

```
12 8
1 1 0 0 0 0 0 0
0 0 1 1 0 0 0 0
0 0 0 0 1 1 0 0
0 0 0 0 0 0 1 1
1 0 1 0 0 0 0 0
0 1 0 1 0 0 0 0
0 0 0 0 1 0 1 0
0 0 0 0 0 1 0 1
1 0 0 0 1 0 0 0
0 1 0 0 0 1 0 0
0 0 1 0 0 0 1 0
0 0 0 1 0 0 0 1
```

The Markov basis associated to the kernel of this matrix can be computed using the command markov no3way, which writes its output to the file no3way.mar. This file is represented in matrix format as:

```
1 8
1 -1 -1 1 -1 1 1 -1
```

The code outputs the Markov basis up to sign. In this case, the Markov basis consists of two elements, the indicated $2 \times 2 \times 2$ table, and its negative. This move would be represented in tableau notation as

$$
\begin{bmatrix} 1 & 1 & 1 \\ 1 & 2 & 2 \\ 2 & 1 & 2 \\ 2 & 2 & 1 \end{bmatrix} - \begin{bmatrix} 1 & 1 & 2 \\ 1 & 2 & 1 \\ 2 & 1 & 1 \\ 2 & 2 & 2 \end{bmatrix}.
$$

The move corresponds to the quartic equation at the end of Example 1.2.7. $\qquad\square$

One of the big challenges in the study of Markov bases of hierarchical models is to find descriptions of the Markov bases as the simplicial complex Γ and the numbers of states of the random variables vary. When it is not possible to give an explicit description of the Markov basis (that is, a list of all types of moves needed in the Markov basis), we might still hope to provide structural or asymptotic information about the types of moves that could arise. In the remainder of this section, we describe some results of this type.

For a simplicial complex Γ, let $\mathcal{G}(\Gamma) = \cup_{S \in \Gamma} S$ denote the *ground set* of Γ.

Definition 1.2.13. A simplicial complex Γ is *reducible*, with reducible decomposition (Γ_1, S, Γ_2) and *separator* $S \subset \mathcal{G}(\Gamma)$, if it satisfies $\Gamma = \Gamma_1 \cup \Gamma_2$ and $\Gamma_1 \cap \Gamma_2 = 2^S$. Furthermore, we here assume that neither Γ_1 nor Γ_2 is 2^S. A simplicial complex is *decomposable* if it is reducible and Γ_1 and Γ_2 are decomposable or simplices (that is, of the form 2^R for some $R \subseteq [m]$).

Of the examples we have seen so far, the simplicial complexes [1][2] and [12][23][345] are decomposable, whereas the simplicial complex [12][13][23] is not reducible. On the other hand, the complex $\Gamma = [12][13][23][345]$ is reducible but not decomposable, with reducible decomposition $([12][13][23], \{3\}, [345])$.

If a simplicial complex has a reducible decomposition, then there is naturally a large class of moves with 1-norm equal to 4 that belong to the lattice $\ker_{\mathbb{Z}} A_\Gamma$. Usually, these moves also appear in some minimal Markov basis.

Lemma 1.2.14. *If Γ is a reducible simplicial complex with reducible decomposition (Γ_1, S, Γ_2), then the following set of moves, represented in tableau notation, belongs to the lattice $\ker_{\mathbb{Z}} A_\Gamma$:*

$$\mathcal{D}(\Gamma_1, \Gamma_2) = \left\{ \begin{bmatrix} i & j & k \\ i' & j & k' \end{bmatrix} - \begin{bmatrix} i & j & k' \\ i' & j & k \end{bmatrix} \; : \; i, i' \in \mathcal{R}_{\mathcal{G}(\Gamma_1) \setminus S}, j \in \mathcal{R}_S, \right.$$

$$\left. k, k' \in \mathcal{R}_{\mathcal{G}(\Gamma_2) \setminus S} \right\}.$$

Theorem 1.2.15 (Markov bases of decomposable models [34, 93]).
If Γ is a decomposable simplicial complex, then the set of moves

$$\mathcal{B} = \bigcup_{(\Gamma_1, S, \Gamma_2)} \mathcal{D}(\Gamma_1, \Gamma_2),$$

with the union over all reducible decompositions of Γ, is a Markov basis for A_Γ.

Example 1.2.16. Consider the 4-chain $\Gamma = [12][23][34]$. This graph has two distinct reducible decompositions with minimal separators, namely $([12], \{2\}, [23][34])$ and $([12][23], \{3\}, [34])$. Therefore, the Markov basis consists of moves of the two types $\mathcal{D}([12], [23][34])$ and $\mathcal{D}([12][23], [34])$, which in tableau notation look like

$$\begin{bmatrix} i_1 & j & i_3 & i_4 \\ i'_1 & j & i'_3 & i'_4 \end{bmatrix} - \begin{bmatrix} i_1 & j & i'_3 & i'_4 \\ i'_1 & j & i_3 & i_4 \end{bmatrix} \quad \text{and} \quad \begin{bmatrix} i_1 & i_2 & j & i_4 \\ i'_1 & i'_2 & j & i'_4 \end{bmatrix} - \begin{bmatrix} i_1 & i_2 & j & i'_4 \\ i'_1 & i'_2 & j & i_4 \end{bmatrix}.$$

Note that the decomposition ($[12][23], \{2,3\}, [23][34]$) is also a valid reducible decomposition of Γ, but it does not produce any new Markov basis elements. $\square$

Theorem 1.2.15 is a special case of a more general result which determines the Markov bases for reducible complexes Γ from the Markov bases of the pieces Γ_1 and Γ_2. For details see the articles [35, 59].

One of the remarkable consequences of Theorem 1.2.15 is that the structure of the Markov basis of a decomposable hierarchical log-linear model does not depend on the number of states of the underlying random variables. In particular, regardless of the sizes $r_1, r_2, \ldots, r_m$, the Markov basis for a decomposable model always consists of moves with 1-norm equal to 4, with a precise and global combinatorial description. The following theorem of De Loera and Onn [29] says that this nice behavior fails, in the worst possible way, already for the simplest non-decomposable model. We fix $\Gamma = [12][13][23]$ and consider $3 \times r_2 \times r_3$ tables, where r_2, r_3 can be arbitrary. De Loera and Onn refer to these as *slim tables*.

Theorem 1.2.17 (Slim tables). *Let* $\Gamma = [12][13][23]$ *be the 3-cycle and let* $v \in \mathbb{Z}^k$ *be any integer vector. Then there exist* $r_2, r_3 \in \mathbb{N}$ *and a coordinate projection* $\pi : \mathbb{Z}^{3 \times r_2 \times r_3} \to \mathbb{Z}^k$ *such that every minimal Markov basis for* Γ *on* $3 \times r_2 \times r_3$ *tables contains a vector* u *such that* $\pi(u) = v$.

In particular, Theorem 1.2.17 shows that there is no hope for a general bound on the 1-norms of Markov basis elements for non-decomposable models, even for a fixed simplicial complex Γ. On the other hand, if only one of the table dimensions is allowed to vary, then there is a bounded finite structure to the Markov bases. This theorem was first proven in [62] and generalizes a result in [81].

Theorem 1.2.18 (Long tables). *Let* Γ *be a simplicial complex and fix* $r_2, \ldots, r_m$. *There exists a number* $b(\Gamma, r_2, \ldots, r_m) < \infty$ *such that the 1-norms of the elements of any minimal Markov basis for* Γ *on* $s \times r_2 \times \cdots \times r_m$ *tables are less than or equal to* $b(\Gamma, r_2, \ldots, r_m)$. *This bound is independent of* s, *which can grow large.*

From Theorem 1.2.15, we saw that if Γ is decomposable and not a simplex, then $b(\Gamma, r_2, \ldots, r_m) = 4$. One of the first discovered results in the non-decomposable case was $b([12][13][23], 3, 3) = 20$, a result obtained by Aoki and Takemura [10]. In general, it seems a difficult problem to actually compute the values $b(\Gamma, r_2, \ldots, r_m)$, although some recent progress was reported by Hemmecke and Nairn [58]. The proof of Theorem 1.2.18 only gives a theoretical upper bound on this quantity, involving other numbers that are also difficult to compute.

1.3 The Many Bases of an Integer Lattice

The goal of this section is to study the notion of a Markov basis in more combinatorial and algebraic detail. In particular, we will explain the relationships between Markov bases and other classical notions of a basis of an integral lattice. In the setting of log-linear models and hierarchical models, this integral lattice would be

$\mathrm{ker}_{\mathbb{Z}}(A)$ as in Definition 1.1.12. One of the highlights of this section is Theorem 1.3.6 which makes a connection between Markov bases and commutative algebra.

We fix any sublattice $\mathcal{L}$ of $\mathbb{Z}^k$ with the property that the only non-negative vector in $\mathcal{L}$ is the origin. In other words, $\mathcal{L}$ is a subgroup of $(\mathbb{Z}^k, +)$ that satisfies

$$\mathcal{L} \cap \mathbb{N}^k = \{0\}.$$

This hypothesis holds for a lattice $\mathrm{ker}_{\mathbb{Z}}(A)$ given by a non-negative integer matrix A, as encountered in the previous sections, and it ensures that the fiber of any point $u \in \mathbb{N}^k$ is a finite set. Here, by the *fiber* of u we mean the set of all non-negative vectors in the same residue class modulo $\mathcal{L}$. This set is denoted by

$$\mathcal{F}(u) := (u + \mathcal{L}) \cap \mathbb{N}^k = \{v \in \mathbb{N}^k : u - v \in \mathcal{L}\}.$$

There are four fundamental problems concerning the fibers: counting $\mathcal{F}(u)$, enumerating $\mathcal{F}(u)$, optimizing over $\mathcal{F}(u)$ and sampling from $\mathcal{F}(u)$.

The optimization problem is the *integer programming problem in lattice form*:

$$\text{minimize } w \cdot v \quad \text{subject to} \quad v \in \mathcal{F}(u). \tag{1.3.1}$$

The sampling problem asks for a random point from $\mathcal{F}(u)$, drawn according to some distribution on $\mathcal{F}(u)$. As seen in Section 1.1, the ability to sample from the hypergeometric distribution is needed for hypothesis testing, but sometimes the uniform distribution is also used [32].

These four problems can be solved if we are able to perform (random) walks that connect the fibers $\mathcal{F}(u)$ using simple steps from the lattice $\mathcal{L}$. To this end, we shall introduce a hierarchy of finite bases in $\mathcal{L}$. The hierarchy looks like this:

$$\text{lattice basis} \subset \text{Markov basis} \subset \text{Gröbner basis}$$
$$\subset \text{universal Gröbner basis} \subset \text{Graver basis}.$$

The purpose of this section is to introduce these five concepts. The formal definitions will be given after the next example. Example 1.3.1 serves as a warm-up, and it shows that all four inclusions among the five different bases can be strict.

Example 1.3.1. Let $k = 4$ and consider the three-dimensional lattice

$$\mathcal{L} = \{(u_1, u_2, u_3, u_4) \in \mathbb{Z}^4 : 3u_1 + 3u_2 + 4u_3 + 5u_4 = 0\}.$$

The following three vectors form a *lattice basis* of $\mathcal{L}$:

$$(1, -1, 0, 0), \ (0, 1, -2, 1), \ (0, 3, -1, -1). \tag{1.3.2}$$

The choice of a lattice basis is not unique, but its cardinality 3 is an invariant of the lattice. Augmenting (1.3.2) by the next vector gives a *Markov basis* of $\mathcal{L}$:

$$(0, 2, 1, -2). \tag{1.3.3}$$

The Markov basis of $\mathcal{L}$ is not unique but it is "more unique" than a lattice basis. The cardinality 4 of the minimal Markov basis is an invariant of the lattice. Augmenting (1.3.2) and (1.3.3) by the following two vectors leads to a *Gröbner basis* of $\mathcal{L}$:

$$(0, 1, 3, -3), \ (0, 0, 5, -4). \tag{1.3.4}$$

This Gröbner basis is *reduced*. The reduced Gröbner basis of a lattice is not unique, but there are only finitely many distinct reduced Gröbner bases. They depend on the choice of a cost vector. Here we took $w = (100, 10, 1, 0)$. This choice ensures that the leftmost non-zero entry in each of our vectors is positive. We note that the cardinality of a reduced Gröbner basis is *not* an invariant of the lattice $\mathcal{L}$.

The *universal Gröbner basis* of a lattice is unique (if we identify each vector with its negative). The universal Gröbner basis of $\mathcal{L}$ consists of 14 vectors. In addition to the six above, it comprises the eight vectors

$$(1, 0, -2, 1), \ (3, 0, -1, -1), \ (2, 0, 1, -2), \ (1, 0, 3, -3),$$
$$(0, 4, -3, 0), \ (4, 0, -3, 0), \ (0, 5, 0, -3), \ (5, 0, 0, -3).$$

Besides the 14 vectors in the universal Gröbner basis, the *Graver basis* of $\mathcal{L}$ contains the following additional ten vectors:

$$(1, 1, 1, -2), \ (1, 2, -1, -1), \ (2, 1, -1, -1), \ (1, 3, -3, 0), \ (2, 2, -3, 0),$$
$$(3, 1, -3, 0), \ (1, 4, 0, -3), \ (2, 3, 0, -3), \ (3, 2, 0, -3), \ (4, 1, 0, -3).$$

The Graver basis of a lattice is unique (up to negating vectors). □

We shall now give precise definitions for the five notions in our hierarchy of bases for an integer lattice $\mathcal{L} \subset \mathbb{Z}^k$. A *lattice basis* is a subset $\mathcal{B} = \{b_1, b_2, \ldots, b_r\}$ of $\mathcal{L}$ such that every vector v in $\mathcal{L}$ has a unique representation

$$v \ = \ \lambda_1 b_1 + \lambda_2 b_2 + \cdots + \lambda_r b_r, \quad \text{with } \lambda_i \in \mathbb{Z}.$$

All lattice bases of $\mathcal{L}$ have the same cardinality r. Each of them specifies a particular isomorphism $\mathcal{L} \simeq \mathbb{Z}^r$. The number r is the *rank* of the lattice $\mathcal{L}$.

Consider an arbitrary finite subset $\mathcal{B}$ of $\mathcal{L}$. This subset determines an undirected graph $\mathcal{F}(u)_\mathcal{B}$ whose nodes are the elements in the fiber $\mathcal{F}(u)$. Two nodes v and v' are connected by an undirected edge in $\mathcal{F}(u)_\mathcal{B}$ if either $v - v'$ or $v' - v$ is in $\mathcal{B}$. We say that $\mathcal{B}$ is a *Markov basis* for $\mathcal{L}$ if the graphs $\mathcal{F}(u)_\mathcal{B}$ are connected for all $u \in \mathbb{N}^k$. (Note that this definition slightly differs from the one used in Sections 1.1 and 1.2, where it was more convenient to include both a vector and its negative in the Markov basis.) We will usually require Markov bases to be minimal with respect to inclusion. With this minimality assumption, *the Markov basis $\mathcal{B}$ is essentially unique*, in the sense made precise in Theorem 1.3.2 below.

Every vector $b \in \mathcal{L}$ can be written uniquely as the difference $b = b^+ - b^-$ of two non-negative vectors with disjoint support. The *fiber of b* is the congruence class of $\mathbb{N}^k$ modulo $\mathcal{L}$ which contains both b^+ and b^-. In symbols,

$$\text{fiber}(b) \ := \ \mathcal{F}(b^+) \ = \ \mathcal{F}(b^-).$$

Theorem 1.3.2. *For a minimal Markov basis $\mathcal{B}$ of a lattice $\mathcal{L}$, the multiset*

$$\{ \operatorname{fiber}(b) : b \in \mathcal{B} \} \tag{1.3.5}$$

is an invariant of the lattice $\mathcal{L} \subset \mathbb{Z}^k$ and hence so is the cardinality of $\mathcal{B}$.

Proof. We shall give an invariant characterization of the multiset (1.3.5). For any fiber $f \in \mathbb{N}^k/\mathcal{L}$ we define a graph G_f as follows. The nodes are the non-negative vectors in $\mathbb{N}^k$ which lie in the congruence class f, and two nodes u and v are connected by an edge if there exists an index i such that $u_i \neq 0$ and $v_i \neq 0$. Equivalently, $\{u, v\}$ is an edge of G_f if and only if $\operatorname{fiber}(u - v) \neq f$.

We introduce the following multiset of fibers:

$$\{ f \in \mathbb{N}^k/\mathcal{L} : \text{ the graph } G_f \text{ is disconnected } \}. \tag{1.3.6}$$

The multiset structure on the underlying set is as follows. The multiplicity of f in (1.3.6) is one less than the number of connected components of the graph G_f.

We claim that the multisets (1.3.5) and (1.3.6) are equal. In proving this claim, we shall use induction on the partially ordered set (poset) $\mathbb{N}^k/\mathcal{L}$. This set inherits its poset structure from the partial order on $\mathbb{N}^k$. Namely, two fibers f and f' satisfy $f' \leq f$ if and only if there exist $u, u' \in \mathbb{N}^k$ such that

$$f = \mathcal{F}(u) \text{ and } f' = \mathcal{F}(u') \text{ and } u' \leq u \text{ (coordinatewise)}.$$

Consider any fiber $f = \mathcal{F}(u)$ and let $C_1, \ldots, C_s$ be the connected components of G_f. Suppose that $\mathcal{B}$ is any minimal Markov basis and consider $\mathcal{B}_f = \{b \in \mathcal{B} : \operatorname{fiber}(b) = f\}$. We will reconstruct all possible choices for $\mathcal{B}_f$. In order to prove the theorem, we must show that each of them has cardinality $s - 1$.

By induction, we may assume that $\mathcal{B}_{f'}$ has already been constructed for all fibers f' which are below f in the poset $\mathbb{N}^k/\mathcal{L}$. Let $\mathcal{B}_{<f}$ be the union of these sets $\mathcal{B}_{f'}$ where $f' < f$. The connected components of the graph $\mathcal{F}(u)_{\mathcal{B}_{<f}}$ are precisely the components $C_1, \ldots, C_s$. The reason is that any two points in the same component C_i can be connected by a sequence of moves from a smaller fiber f', but no point in C_i can be connected to a point in a different component C_j by such moves. Therefore, all the possible choices for $\mathcal{B}_f$ are obtained as follows. First we fix a spanning tree on the components $C_1, \ldots, C_s$. Second, for any edge $\{C_i, C_j\}$ in that spanning tree, we pick a pair of points $u \in C_i$ and $v \in C_j$. Finally, the desired set $\mathcal{B}_f$ consists of the resulting $s - 1$ difference vectors $u - v$. This proves $\#\mathcal{B}_f = s - 1$, as desired.   □

The previous proof gives a purely combinatorial algorithm which constructs the minimal Markov basis of a lattice $\mathcal{L}$. We fix a total order on the set of fibers $\mathbb{N}^k/\mathcal{L}$ which refines the natural partial order. Starting with the first fiber $f = \mathcal{F}(0) = \{0\}$ and the empty partial Markov basis $\mathcal{B}_{<0} = \emptyset$, we consider an arbitrary fiber f and the already computed partial Markov basis $\mathcal{B}_{<f}$. The steps of the algorithm are now exactly as in the proof:

1. Identify the connected components $C_1, \ldots, C_s$ of the graph G_f.

2. Pick a spanning tree on $C_1, \ldots, C_s$.

3. For any edge $\{C_i, C_j\}$ of the tree, pick points $u \in C_i$ and $v \in C_j$.

4. Define $\mathcal{B}_f$ as the set of those $s - 1$ difference vectors $u - v$.

5. Move on to the next fiber (unless you are sure to be done).

Example 1.3.3. We demonstrate how this method works for the lattice in Example 1.3.1. Recall that $\mathcal{L}$ is the kernel of the linear map

$$\pi : \mathbb{Z}^4 \to \mathbb{Z}, \quad (u_1, u_2, u_3, u_4) \mapsto 3u_1 + 3u_2 + 4u_3 + 5u_4.$$

The poset of fibers is a subposet of the poset of non-negative integers:

$$\mathbb{N}^4 / \mathcal{L} \;=\; \pi(\mathbb{N}^4) \;=\; \{0, 3, 4, 5, 6, \ldots\} \;\subset\; \mathbb{N}.$$

The fiber 0 is trivial, so our algorithm starts with $f = 3$ and $\mathcal{B}_{<3} = \emptyset$. The graph G_3 has two connected components

$$C_1 \;=\; \{(1,0,0,0)\} \quad \text{and} \quad C_2 \;=\; \{(0,1,0,0)\},$$

so we have no choice but to take $\mathcal{B}_3 = \{(1, -1, 0, 0)\}$. The next steps are:

- G_4 has only one node $(0,0,1,0)$ hence $\mathcal{B}_4 = \emptyset$.

- G_5 has only one node $(0,0,0,1)$ hence $\mathcal{B}_5 = \emptyset$.

- $G_6 = \{(2,0,0,0), (1,1,0,0), (0,2,0,0)\}$ is connected hence $\mathcal{B}_6 = \emptyset$.

- $G_7 = \{(1,0,1,0), (0,1,1,0)\}$ is connected hence $\mathcal{B}_7 = \emptyset$.

- G_8 has two connected components, $C_1 = \{(1,0,0,1), (0,1,0,1)\}$ and $C_2 = \{(0,0,2,0)\}$, and we decide to take $\mathcal{B}_8 = \{(0,1,-2,1)\}$.

- G_9 has two connected components, namely $C_1 = \{(3,0,0,0), (2,1,0,0), (1,2,0,0), (0,3,0,0)\}$ and $C_2 = \{(0,0,1,1)\}$. We take $\mathcal{B}_9 = \{(0,3,-1,-1)\}$.

- G_{10} has two connected components, $C_1 = \{(2,0,1,0), (1,1,1,0), (0,2,1,0)\}$ and $C_2 = \{(0,0,0,2)\}$, and we take $\mathcal{B}_{10} = \{(0,2,1,-2)\}$.

At this stage, divine inspiration tells us that the Markov basis for $\mathcal{L}$ is already complete. So, we decide to stop and we output $\mathcal{B} = \mathcal{B}_{\leq 10}$. The multiset of Markov fibers (1.3.5) is the set $\{3, 8, 9, 10\}$, where each element has multiplicity 1. $\quad\square$

There are two obvious problems with this algorithm. The first is that we need a termination criterion, and the second concerns the combinatorial explosion (which becomes serious for $n - \mathrm{rank}(\mathcal{L}) \geq 3$) of having to look at many fibers until a termination criterion kicks in. The first problem can be addressed by deriving a general bound on the sizes of the coordinates of any element in the Graver basis of

$\mathcal{L}$. Such a bound is given in [87, Theorem 4.7, p. 33]. However, a more conceptual solution for both problems can be given by recasting the Markov basis property in terms of commutative algebra [25, 87]. This will be done in Theorem 1.3.6 below.

First, however, we shall define the other three bases of $\mathcal{L}$. Fix a generic cost vector $w \in \mathbb{R}^k$. Here *generic* means that each integer program (1.3.1) has only one optimal solution. Suppose that $b \cdot w < 0$ for all $b \in \mathcal{B}$. We regard $\mathcal{F}(u)_{\mathcal{B}}$ as a directed graph by introducing a directed edge $v \to v'$ whenever $v' - v$ is in $\mathcal{B}$. In this manner, $\mathcal{F}(u)_{\mathcal{B}}$ becomes an acyclic directed graph. We say that $\mathcal{B}$ is a *Gröbner basis* of $\mathcal{L}$ if the directed graph $\mathcal{F}(u)_{\mathcal{B}}$ has a unique sink, for all $u \in \mathbb{N}^k$.

Remark 1.3.4. *If $\mathcal{B}$ is a Gröbner basis then the sink of the directed graph $\mathcal{F}(u)_{\mathcal{B}}$ is the optimal solution of the integer programming problem (1.3.1). For more background on the use of Gröbner bases in integer programming we refer to [87, §5].*

Among all Gröbner bases for $\mathcal{L}$ there is a distinguished *reduced Gröbner basis* which is unique when w is fixed. It consists of all vectors $b \in \mathcal{L}$ such that b^- is a sink (in its own fiber), b^+ is not a sink, but $b^+ - e_i$ is a sink for all i with $b_i > 0$.

It is known that there are only finitely many distinct reduced Gröbner bases, as w ranges over generic vectors in $\mathbb{R}^k$. The union of all reduced Gröbner bases is the *universal Gröbner basis* of $\mathcal{L}$.

All of the bases of $\mathcal{L}$ discussed so far are contained in the *Graver basis*. The Graver basis $\mathcal{G}$ of our lattice $\mathcal{L}$ is defined as follows. Fix a sign vector $\sigma \in \{-1, +1\}^k$ and consider the semigroup

$$\mathcal{L}_\sigma \quad := \quad \{\, v \in \mathcal{L} \ : \ v_i \cdot \sigma_i \geq 0 \,\}.$$

This semigroup has a unique minimal finite generating set $\mathcal{G}_\sigma$ called the *Hilbert basis* of $\mathcal{L}_\sigma$. The *Graver basis* $\mathcal{G}$ of $\mathcal{L}$ is the union of these Hilbert bases:

$$\mathcal{G} \quad := \quad \bigcup_{\sigma \in \{-1,+1\}^k} \mathcal{G}_\sigma.$$

This set is finite because each of the Hilbert bases $\mathcal{G}_\sigma$ is finite.

Proposition 1.3.5. *The Graver basis $\mathcal{G}$ is the unique minimal subset of the lattice $\mathcal{L}$ such that every vector $v \in \mathcal{L}$ has a sign-consistent representation in terms of $\mathcal{G}$:*

$$v = \sum_{g \in \mathcal{G}} \lambda_g \cdot g \quad \text{with } \lambda_g \in \mathbb{N} \quad \text{and} \quad |v_i| = \sum_{g \in \mathcal{G}} \lambda_g \cdot |g_i| \quad \text{for all } i \in [k].$$

Markov bases, Gröbner bases, Hilbert bases, and Graver bases of integer lattices can be computed using the software 4ti2, which was developed by Raymond Hemmecke and his collaborators [57]. Further computations with 4ti2 will be shown in the exercises in Chapter 6.

We now come to the interpretation of our bases in terms of algebraic geometry. The given lattice $\mathcal{L} \subset \mathbb{Z}^k$ is represented by the corresponding *lattice ideal*

$$I_\mathcal{L} \quad := \quad \langle\, p^u - p^v \ : \ u, v \in \mathbb{N}^k \text{ and } u - v \in \mathcal{L} \,\rangle \quad \subset \quad \mathbb{R}[p_1, p_2, \ldots, p_k].$$

Here $p_1, \ldots, p_k$ are indeterminates, and $p^u = p_1^{u_1} p_2^{u_2} \cdots p_k^{u_k}$ denotes monomials in these indeterminates. In our applications, p_i will represent the probability of observing the ith state of a random variable with k states. Hilbert's Basis Theorem states that every ideal in the polynomial ring $\mathbb{R}[p_1, p_2, \ldots, p_k]$ is finitely generated. The finiteness of Markov bases is thus implied by the following result of [33], which was one of the starting points for the field of algebraic statistics.

Theorem 1.3.6 (Fundamental theorem of Markov bases). *A subset $\mathcal{B}$ of the lattice $\mathcal{L}$ is a Markov basis if and only if the corresponding set of binomials $\{ p^{b^+} - p^{b^-} : b \in \mathcal{B} \}$ generates the lattice ideal $I_{\mathcal{L}}$.*

The notions of Gröbner bases and Graver bases are also derived from their algebraic analogues. For a detailed account see [87]. In that book, as well as in most statistical applications, the lattice $\mathcal{L}$ arises as the kernel of an integer matrix A. The algebraic theory for arbitrary lattices is found in [70, Chapter 7]. The multiset in Theorem 1.3.2 corresponds to the multidegrees of the minimal generators of $I_{\mathcal{L}}$.

Let $A = (a_{ij}) \in \mathbb{N}^{d \times k}$ be a non-negative integer matrix. We assume that all the column sums of A are equal. The columns $a_j = (a_{1j}, a_{2j}, \ldots, a_{dj})^T$ of A represent monomials $\theta^{a_j} = \theta_1^{a_{1j}} \theta_2^{a_{2j}} \cdots \theta_d^{a_{dj}}$ in auxiliary unknowns θ_i that correspond to model parameters. The monomials θ^{a_j} all have the same degree.

The matrix A determines a monomial map

$$\phi_A : \mathbb{C}^d \to \mathbb{C}^k, \ \theta \mapsto (\theta^{a_1}, \theta^{a_2}, \ldots, \theta^{a_k}).$$

The closure of the image of this map is the *affine toric variety* V_A associated to the matrix A. The connection to tori arises from the fact that V_A is the closure of the image of the algebraic torus $\phi_A((\mathbb{C}^*)^d)$. If we restrict the map ϕ_A to the positive reals $\mathbb{R}_{>0}^d$, and consider the image in the probability simplex $\Delta_{k-1} = \mathbb{R}_{\geq 0}^k / \text{scaling}$, we get the log-linear model $\mathcal{M}_A$. For this reason, log-linear models are sometimes known as *toric models*. See Section 1.2 in [73] for more on toric models.

More generally, a *variety* is the solution set to a simultaneous system of polynomial equations. If I is an ideal, then $V(I)$ is the variety defined by the vanishing of all polynomials in I. Often, we might need to be more explicit about *where* the solutions to this system of equations lie, in which case we use the notation $V_*(I)$ to denote the solutions constrained by condition $*$. The different types of solution spaces will be illustrated in Example 1.3.8.

Proposition 1.3.7. *The lattice ideal $I_{\mathcal{L}}$ for $\mathcal{L} = \ker_{\mathbb{Z}}(A)$ is a prime ideal. Its homogeneous elements are exactly the homogeneous polynomials in $\mathbb{R}[p_1, \ldots, p_k]$ that vanish on probability distributions in the log-linear model specified by the matrix A. In other words, the toric variety $V_A = V(I_{\mathcal{L}})$ is the Zariski closure of the log-linear model $\mathcal{M}_A$.*

The binomials corresponding to the Markov basis generate the ideal $I_{\mathcal{L}}$ and hence they cut out the toric variety $V_A = V(I_{\mathcal{L}})$. However, often one does not need the full Markov basis to define the toric variety set-theoretically. Finding good choices of such partial bases is a delicate matter, as the following example demonstrates.

Example 1.3.8. Let $d = 3$, $k = 9$ and consider the matrix

$$
A \;=\;
\begin{matrix}
 & p_1 & p_2 & p_3 & p_4 & p_5 & p_6 & p_7 & p_8 & p_9 \\
 & \begin{pmatrix} 3 & 0 & 0 & 2 & 1 & 2 & 1 & 0 & 0 \\ 0 & 3 & 0 & 1 & 2 & 0 & 0 & 2 & 1 \\ 0 & 0 & 3 & 0 & 0 & 1 & 2 & 1 & 2 \end{pmatrix} &&&&&&&&
\end{matrix}
\qquad (1.3.7)
$$

and the associated monomial parametrization

$$
\phi_A : (\theta_1, \theta_2, \theta_3) \mapsto (\theta_1^3, \theta_2^3, \theta_3^3, \theta_1^2\theta_2, \theta_1\theta_2^2, \theta_1^2\theta_3, \theta_1\theta_3^2, \theta_2^2\theta_3, \theta_2\theta_3^2). \qquad (1.3.8)
$$

The minimal Markov basis of the lattice $\mathcal{L} = \mathrm{ker}_{\mathbb{Z}}(A)$ consists of 17 vectors. These vectors correspond to the set of all 17 quadratic binomials listed in (1.3.9), (1.3.10), (1.3.11) and (1.3.12) below. We start out with the following six binomials:

$$
\{\, p_1 p_5 - p_4^2,\ p_2 p_4 - p_5^2,\ p_1 p_7 - p_6^2,\ p_3 p_6 - p_7^2,\ p_2 p_9 - p_8^2,\ p_3 p_8 - p_9^2 \,\}. \qquad (1.3.9)
$$

The vectors corresponding to (1.3.9) form a basis for the kernel of A as a vector space over the rational numbers $\mathbb{Q}$ but they do not span $\mathcal{L}$ as a lattice over $\mathbb{Z}$. Nevertheless, a positive vector $p = (p_1, \ldots, p_9)$ is a common zero of these six binomials if and only if p lies in the image of a positive vector $(\theta_1, \theta_2, \theta_3)$ under the map ϕ_A. The same statement fails badly for non-negative vectors. Namely, in addition to $V_{\geq 0}(I_{\mathcal{L}})$, which is the closure of the log-linear model, the non-negative variety of (1.3.9) has seven extraneous components, which are not in the closure of the log-linear model $\mathcal{M}_A$. One such component is the three-dimensional orthant

$$
\{\, (p_1, p_2, p_3, 0, 0, 0, 0, 0, 0) \,:\, p_1, p_2, p_3 \in \mathbb{R}_{\geq 0} \,\} \quad \subset \quad V_{\geq 0}((1.3.9)).
$$

We invite the reader to find the six others. These seven extraneous components disappear again if we augment (1.3.9) by the following three binomials:

$$
\{\, p_1 p_2 - p_4 p_5,\ p_1 p_3 - p_6 p_7,\ p_2 p_3 - p_8 p_9 \,\}. \qquad (1.3.10)
$$

Hence the non-negative variety defined by the nine binomials in (1.3.9) and (1.3.10) is the closure of the log-linear model. The same holds over the reals:

$$
V_{\geq 0}(I_{\mathcal{L}}) \;=\; V_{\geq 0}((1.3.9),\,(1.3.10)) \quad \text{and} \quad V_{\mathbb{R}}(I_{\mathcal{L}}) \;=\; V_{\mathbb{R}}((1.3.9),\,(1.3.10)).
$$

On the other hand, the varieties over the complex numbers are still different:

$$
V_{\mathbb{C}}(I_{\mathcal{L}}) \;\neq\; V_{\mathbb{C}}((1.3.9),\,(1.3.10)).
$$

The complex variety of the binomials in (1.3.9) and (1.3.10) breaks into three irreducible components, each of which is a multiplicative translate of the toric variety $V_{\mathbb{C}}(I_{\mathcal{L}})$. Namely, if we start with any point p in $V_{\mathbb{C}}(I_{\mathcal{L}})$ and we replace p_4 by ηp_4 and p_5 by $\eta^2 p_5$, where $\eta = -\frac{1}{2} + \frac{\sqrt{3}}{2}i$ is a primitive cube root of unity, then the new vector is no longer in $V_{\mathbb{C}}(I_{\mathcal{L}})$ but still satisfies the nine binomials in (1.3.9) and (1.3.10). This is detected algebraically as follows. The binomial

$$p_1^3 p_8^3 - p_5^3 p_6^3 \;=\; (p_1 p_8 - p_5 p_6)(p_1 p_8 - \eta p_5 p_6)(p_1 p_8 - \eta^2 p_5 p_6)$$

lies in the ideal of (1.3.9) and (1.3.10) but none of its factors does. To remove the two extraneous complex components, we add six more binomials:

$$\big\{\, p_1 p_8 - p_5 p_6,\; p_1 p_9 - p_4 p_7,\; p_2 p_6 - p_4 p_8,\; p_2 p_7 - p_5 p_9,$$
$$p_3 p_4 - p_6 p_9,\; p_3 p_5 - p_7 p_8 \,\big\}. \quad (1.3.11)$$

Let J denote the ideal generated by the 15 binomials in (1.3.9), (1.3.10) and (1.3.11). The radical of the ideal J equals $I_{\mathcal{L}}$. This means that the complex variety of J coincides with $V_{\mathbb{C}}(I_{\mathcal{L}})$. However, the ideal J is still strictly contained in $I_{\mathcal{L}}$. To get the Markov basis, we still need to add the following two binomials:

$$\big\{\, p_6 p_8 - p_4 p_9,\; p_5 p_7 - p_4 p_9 \,\big\}. \quad (1.3.12)$$

The lattice $\mathcal{L}$ in this example has the following special property. Its Markov basis consists of quadratic binomials, but no Gröbner basis of $I_{\mathcal{L}}$ has only quadratic elements. Using the software Gfan [63], one can easily check that $\mathcal{L}$ has precisely $54,828$ distinct reduced Gröbner bases. Each of them contains at least one binomial of degree 3. For instance, the reduced Gröbner basis with respect to the reverse lexicographic order consists of our 17 quadrics and the two cubics $p_1 p_7 p_8 - p_4 p_6 p_9$ and $p_7^2 p_8 - p_6 p_9^2$. $\qquad \square$

We remark that we will see quadratic binomials of the form $p_i p_j - p_k p_l$ again in Chapter 3, where they naturally correspond to conditional independence relations. The ideal of such relations will make its first appearance in Definition 3.1.5. We close the current chapter by describing a simple log-linear model in which the algebraic structure from Example 1.3.8 arises.

Example 1.3.9. Bobby and Sally play *Rock-Paper-Scissors* according to the following rules. One round consists of three games, and it is not permissible to make three different choices in one round. Should this happen then the round of three games is repeated until a valid outcome occurs. After $n = 1000$ rounds of playing, Sally decides to analyze Bobby's choices that can be summarized in the vector

$$u \;=\; \big(u_{rrr}, u_{ppp}, u_{sss}, u_{rrp}, u_{rpp}, u_{rrs}, u_{rss}, u_{pps}, u_{pss}\big),$$

where u_{rrr} is the number of rounds in which Bobby picks rock three times, u_{rrp} is the number of rounds in which he picks rock twice and paper once, and so on. Sally suspects that Bobby makes independent random choices picking rock with probability θ_1, paper with probability θ_2, and scissors with probability $\theta_3 = 1 - \theta_1 - \theta_2$. Let p_{rrr}, p_{ppp}, etc. be the probabilities of Bobby's choices. Under the hypothesis of random choices, the vector of rescaled probabilities

$$(3p_{rrr}, 3p_{ppp}, 3p_{sss}, p_{rrp}, p_{rpp}, p_{rrs}, p_{rss}, p_{pps}, p_{pss})$$

is a point in the toric variety discussed in Example 1.3.8. Sally can thus use the Markov basis given there to test her hypothesis that Bobby makes random choices. All she needs to do is to run the Metropolis-Hastings Algorithm 1.1.13, and then apply the hypothesis testing framework that was outlined in Section 1.1. Note, however, that the rescaling of the probabilities leads to an adjustment of the hypergeometric distribution in (1.1.10). In this adjustment we divide the numerator of (1.1.10) by $3^{u_{rrr}+u_{ppp}+u_{sss}}$ (or multiply by $3^{u_{rrp}+u_{rpp}+u_{rrs}+u_{rss}+u_{pps}+u_{pss}}$) and apply the corresponding division (or multiplication) to each term in the sum in the denominator.   □

Chapter 2

Likelihood Inference

This chapter is devoted to algebraic aspects of maximum likelihood estimation and likelihood ratio tests. Both of these statistical techniques rely on maximization of the likelihood function, which maps the parameters indexing the probability distributions in a statistical model to the likelihood of observing the data at hand. Algebra enters the playing field in two ways. First, computing maximum likelihood (ML) estimates often requires solving algebraic critical equations. Second, many models can be described as semi-algebraic subsets of the parameter space of a nice ambient model. In that setting, algebraic techniques are helpful for determining the behavior of statistical procedures such as the likelihood ratio test.

Section 2.1 begins with a discussion of the computation of maximum likelihood estimates in discrete models, including the log-linear models encountered in Chapter 1. The second half of Section 2.1 introduces Gaussian models. Section 2.2 presents algebraic techniques for the computation of maximum likelihood estimates for discrete models that are defined implicitly, by polynomial equations in the probabilities. In Section 2.3, we turn to likelihood ratio tests, which constitute a general approach to solving hypothesis testing problems. A crucial ingredient to this methodology is asymptotic distribution theory for large sample size, and we discuss how the geometry of the parameter space affects the asymptotics.

2.1 Discrete and Gaussian Models

Let $\mathcal{P}_\Theta = \{P_\theta : \theta \in \Theta\}$ be a statistical model with finite dimensional open parameter space $\Theta \subseteq \mathbb{R}^d$. We assume throughout that each distribution P_θ has a density function $p_\theta(x)$ with respect to some fixed measure ν. In other words, $P_\theta(A) = \int_A p_\theta(x)d\nu(x)$ for all measurable sets A.

Let $X^{(1)}, X^{(2)}, \ldots, X^{(n)} \sim P_\theta$ be independent random vectors that are identically distributed according to some unknown probability distribution $P_\theta \in \mathcal{P}_\Theta$.

The *likelihood function* is the function

$$L_n(\theta) = \prod_{i=1}^{n} p_\theta(X^{(i)}),$$

and the *log-likelihood function* is $\ell_n(\theta) = \log L_n(\theta)$. Often, we will write $L(\theta)$ and $\ell(\theta)$ when the dependence on the sample size n is not important.

Definition 2.1.1. The *maximum likelihood estimator* (ML estimator) of the unknown parameter θ is the random variable

$$\hat{\theta} = \operatorname{argmax}_{\theta \in \Theta} \ell_n(\theta).$$

The *maximum likelihood estimate* of θ for the data $x^{(1)}, \ldots, x^{(n)}$ is the realization of $\hat{\theta}$ obtained by the substitution $X^{(1)} = x^{(1)}, \ldots, X^{(n)} = x^{(n)}$.

   The ML estimator $\hat{\theta}$ is the parameter value such that under the corresponding distribution the likelihood of observing $X^{(1)}, \ldots, X^{(n)}$ is maximal. When $X^{(1)}, \ldots, X^{(n)}$ are discrete random vectors, the density $p_\theta(x)$ determines probabilities, and $\hat{\theta}$ simply maximizes the probability of observing the data.

   Classical results of probability theory guarantee that if the statistical model $\mathcal{P}_\Theta$ satisfies suitable regularity conditions, then $\hat{\theta}$ is an asymptotically unbiased estimator of the true parameter θ and has an asymptotic normal distribution as $n \to \infty$; compare [94]. This asymptotic distribution theory of maximum likelihood estimators parallels the theory of likelihood ratio tests that we will discuss in Section 2.3. In that section, we will see how the asymptotics are related to the geometry of parameter spaces.

   The first place where algebra enters into likelihood theory is the computation of maximum likelihood estimates. In many circumstances this computation amounts to solving an algebraic optimization problem. Indeed, if it happens that

$$\log p_\theta(x) = \log q_1(\theta) + q_2(\theta) \tag{2.1.1}$$

where $q_i \in \mathbb{Q}(\theta)$ are rational functions of θ, then the maximum likelihood estimate $\hat{\theta}$, if it exists, is the solution to a simultaneous system of algebraic equations in θ. These equations are called the *likelihood equations* or *score equations*.

Discrete Models.   One ubiquitous situation where the condition of (2.1.1) is satisfied is for parametric discrete statistical models. In this setting, Θ is an open subset of $\mathbb{R}^d$ (usually the interior of a polyhedron) and we have a rational parametrization map $g : \Theta \to \Delta_{k-1}$, the probability simplex. That is, each coordinate g_i is a rational function in θ, with rational coefficients. Thus, in the discrete case, the maximum likelihood estimation problem amounts to maximizing the function

$$\ell(\theta) = \sum_{i=1}^{k} u_i \log g_i(\theta),$$

where $u_i = \#\{j : X^{(j)} = i\}$ are the coordinates in a vector or table of counts. The likelihood equations in this discrete setting are the following d equations

$$\sum_{i=1}^{k} \frac{u_i}{g_i} \cdot \frac{\partial g_i}{\partial \theta_j} = 0 \qquad \text{for } j = 1, \ldots, d. \tag{2.1.2}$$

Note that the left hand side of each equation is a rational function in d unknowns.

Example 2.1.2 (Independence model). The parametrization of the independence model $\mathcal{M}_{X \perp\!\!\!\perp Y}$ is the map

$$g : \Delta_{r-1} \times \Delta_{c-1} \to \Delta_{rc-1},$$
$$g_{ij}(\alpha, \beta) = \alpha_i \beta_j,$$

where $\alpha_r = 1 - \sum_{i=1}^{r-1} \alpha_i$ and $\beta_c = 1 - \sum_{j=1}^{c-1} \beta_j$. (Recall Definition 1.1.3.) The data is summarized in a table of counts $u \in \mathbb{N}^{r \times c}$. The log-likelihood function is

$$\ell(\alpha, \beta) = \sum_{i,j} u_{ij} \log(\alpha_i \beta_j) = \sum_i u_{i+} \log \alpha_i + \sum_j u_{+j} \log \beta_j,$$

where $u_{i+} = \sum_{j=1}^{c} u_{ij}$ and $u_{+j} = \sum_{i=1}^{c} u_{ij}$. The likelihood equations are

$$\frac{u_{i+}}{\alpha_i} - \frac{u_{r+}}{1 - \sum_{k=1}^{r-1} \alpha_k} = 0,$$

$$\frac{u_{+j}}{\beta_j} - \frac{u_{+c}}{1 - \sum_{k=1}^{c-1} \beta_k} = 0.$$

Clearing denominators and solving the resulting linear systems for α and β yields

$$\hat{\alpha}_i = \frac{u_{i+}}{u_{++}} \qquad \text{and} \qquad \hat{\beta}_j = \frac{u_{+j}}{u_{++}}.$$

These then determine the table of expected counts with entries $\hat{u}_{ij} = n\hat{\alpha}_i\hat{\beta}_j$ that appeared in the chi-square statistic discussed in Section 1.1. $\square$

The process of "clearing denominators", which we cavalierly used in the previous example, can lead to serious difficulties when dealing with models more complicated than the independence model. For a general parametric statistical model for discrete random variables, the likelihood equations have the form of (2.1.2). Here, clearing denominators gives the system of equations

$$\sum_{i=1}^{k} u_i \cdot g_1 \cdots \hat{g}_i \cdots g_k \cdot \frac{\partial g_i}{\partial \theta_j} = 0, \quad j = 1, \ldots, d, \tag{2.1.3}$$

where $\hat{g}_i$ denotes that the i-th element of the product is omitted. Suppose that θ is a parameter vector such that $g_i(\theta) = g_j(\theta) = 0$. Then θ is a solution to (2.1.3) that is not a solution to the original likelihood equations. In particular, if the g_i are generic polynomials, then the solutions to the system (2.1.3) contain a variety of codimension 2 which consists of extraneous solutions to the likelihood equations. While the rational functions that arise in statistics are rarely generic, the introduction of extraneous solutions remains a considerable problem. To illustrate this point, consider the following family of censored exponential distributions.

Example 2.1.3 (Random censoring). We consider families of discrete random variables that arise from randomly censoring exponential random variables. This example describes a special case of a censored continuous time conjunctive Bayesian network; see [15] for more details and derivations.

A random variable T is *exponentially distributed* with rate parameter $\lambda > 0$ if it has the (Lebesgue) density function

$$f(t) \; = \; \lambda \exp(-\lambda t) \cdot 1_{\{t \geq 0\}}, \quad t \in \mathbb{R}.$$

Let T_1, T_2, T_s be independent exponentially distributed random variables with rate parameters $\lambda_1, \lambda_2, \lambda_s$, respectively. Suppose that instead of observing the times T_1, T_2, T_s directly, we can only observe for $i \in \{1, 2\}$ whether T_i occurs before or after T_s. In other words, we observe a discrete random variable with the four states $\emptyset, \{1\}, \{2\}$ and $\{1, 2\}$, which are the elements $i \in \{1, 2\}$ such that $T_i \leq T_s$. This induces a rational map g from the parameter space $(0, \infty)^3$ into the probability simplex Δ_3. The coordinates of g are the functions

$$g_\emptyset(\lambda_1, \lambda_2, \lambda_s) \; = \; \frac{\lambda_s}{\lambda_1 + \lambda_2 + \lambda_s},$$

$$g_{\{1\}}(\lambda_1, \lambda_2, \lambda_s) \; = \; \frac{\lambda_1}{\lambda_1 + \lambda_2 + \lambda_s} \cdot \frac{\lambda_s}{\lambda_2 + \lambda_s},$$

$$g_{\{2\}}(\lambda_1, \lambda_2, \lambda_s) \; = \; \frac{\lambda_2}{\lambda_1 + \lambda_2 + \lambda_s} \cdot \frac{\lambda_s}{\lambda_1 + \lambda_s},$$

$$g_{\{1,2\}}(\lambda_1, \lambda_2, \lambda_s) \; = \; \frac{\lambda_1}{\lambda_1 + \lambda_s} \cdot \frac{\lambda_2}{\lambda_2 + \lambda_s} \cdot \frac{\lambda_1 + \lambda_2 + 2\lambda_s}{\lambda_1 + \lambda_2 + \lambda_s},$$

where $g_\emptyset(\lambda)$ is the probability that $T_1 > T_s$ and $T_2 > T_s$, and so on. Given counts $u_0, u_1, u_2,$ and u_{12}, the log-likelihood function is

$$
\begin{aligned}
\ell(\lambda) \; = \; & (u_1 + u_{12}) \log \lambda_1 + (u_2 + u_{12}) \log \lambda_2 + (u_0 + u_1 + u_2) \log \lambda_s \\
& + u_{12} \log(\lambda_1 + \lambda_2 + 2\lambda_s), \\
& - (u_2 + u_{12}) \log(\lambda_1 + \lambda_s) - (u_1 + u_{12}) \log(\lambda_2 + \lambda_s) \\
& - (u_0 + u_1 + u_2 + u_{12}) \log(\lambda_1 + \lambda_2 + \lambda_s).
\end{aligned}
$$

Since the parametrization involves rational functions of degree zero, we can set $\lambda_s = 1$, and then solve the likelihood equations in λ_1 and λ_2. These likelihood equations are

$$\frac{u_1 + u_{12}}{\lambda_1} + \frac{u_{12}}{\lambda_1 + \lambda_2 + 2} - \frac{u_2 + u_{12}}{\lambda_1 + 1} - \frac{u_0 + u_1 + u_2 + u_{12}}{\lambda_1 + \lambda_2 + 1} = 0$$

$$\frac{u_2 + u_{12}}{\lambda_2} + \frac{u_{12}}{\lambda_1 + \lambda_2 + 2} - \frac{u_1 + u_{12}}{\lambda_2 + 1} - \frac{u_0 + u_1 + u_2 + u_{12}}{\lambda_1 + \lambda_2 + 1} = 0.$$

In this case, clearing denominators always introduces three extraneous solutions $(\lambda_1, \lambda_2) = (-1, -1), (0, -1), (-1, 0)$. The following code for the software Singular exemplifies how to compute all solutions to the equations with the denominators cleared, and how to remove the extraneous solutions via the command sat. The particular counts used for illustration are specified in the third line.

```
LIB "solve.lib";
ring R = 0,(l1,l2),dp;
int u0 = 3; int u1 = 5; int u2 = 7; int u12 = 11;
ideal I = (u1+u12)*(l1+l2+2)*(l1+1)*(l1+l2+1) +
            (u12)*l1*(l1+1)*(l1+l2+1) -
            (u2+u12)*l1*(l1+l2+2)*(l1+l2+1) -
            (u0+u1+u2+u12)*l1*(l1+l2+2)*(l1+1),
          (u2+u12)*(l1+l2+2)*(l2+1)*(l1+l2+1) +
            (u12)*l2*(l2+1)*(l1+l2+1) -
            (u1+u12)*l2*(l1+l2+2)*(l1+l2+1) -
            (u0+u1+u2+u12)*l2*(l1+l2+2)*(l2+1);
ideal K = l1*l2*(l1+1)*(l2+1)*(l1+l2+1)*(l1+l2+2);
ideal J = sat(I,K)[1];
solve(J);
```

In particular, there are three solutions to the likelihood equations, and the maximum likelihood estimate is a degree 3 algebraic function of the data. In general, the maximum likelihood estimate of λ_2 is a root of the cubic polynomial

$$\begin{aligned}
f(\lambda_2) =\ & (u_0 + 2u_1)(u_1 - u_2)\lambda_2^3 \\
& + (u_0 u_1 + u_1^2 - 3u_0 u_2 - 6u_1 u_2 + u_2^2 - 2u_0 u_{12} - 5u_1 u_{12} + u_2 u_{12})\lambda_2^2 \\
& + (2u_0 + 3u_1 - 3u_2 - 3u_{12})(u_2 + u_{12})\lambda_2 \\
& + 2(u_2 + u_{12})^2.
\end{aligned}$$

This polynomial was computed using a variation of the above code, working over the field $\mathbb{Q}(u_0, u_1, u_2, u_{12})$, that can be defined in Singular by

```
ring R = (0,u0,u1,u2,u12),(l1,l2),dp;
```

The leading coefficient reveals that the degree of $f(\lambda_2)$ is 3 for all non-zero vectors of counts u away from the hyperplane defined by $u_1 = u_2$. □

An important step in the above calculation in Singular is the removal of extraneous solutions via the command sat, which computes a saturation ideal. Suppose that I is the ideal generated by the d likelihood equations, with denominators cleared. Let h be the product of all polynomials appearing in the denominators in the rational equations (2.1.2). The solution set of the *saturation ideal*

$$(I : h^\infty) := \{f \in \mathbb{R}[\theta] \ : \ fh^i \in I \text{ for some } i \in \mathbb{N}\}$$

is precisely the closure of those points that are actual complex critical points of the likelihood equations. In the typical situation that the number of complex critical points is finite, saturation eliminates all extraneous solutions. Naive clearing of denominators almost never works in algebraic statistics. The correct way to clear denominators is to pass from I to the saturation ideal $(I : h^\infty)$.

A basic principle of algebraic geometry is that the number of solutions of a system of polynomial or rational equations that depends rationally on parameters is constant except on an algebraic subset of parameter space. In our case, the rational equations under investigation are the likelihood equations and the "varying parameters" are the data. This leads to the following definition.

Definition 2.1.4. The *maximum likelihood degree* (ML degree) of a statistical model satisfying condition (2.1.1) is the number of complex solutions to the likelihood equations for generic data. For discrete random variables, this is the number of complex solutions to the likelihood equations (2.1.2) for generic data u.

The notion of *generic data* means that the number of complex solutions to the equations (2.1.2) is a constant for all data, except possibly for a subset that lies on a lower-dimensional algebraic subset of the data space. We encountered this in Example 2.1.3, where the ML degree is 3 but examining the final cubic equation revealed that the number of critical points will drop to two in the event that $u_1 = u_2$. Such symmetric data is not generic for that problem. As another example, the ML degree of the model of independence is 1.

Having maximum likelihood degree one can be expressed equivalently by saying that the ML estimate is a rational function of the data. As we saw in Section 1.2, the independence model is a special case of a more general family of models with especially nice properties, namely, the decomposable hierarchical models. As explained next, the property of having ML degree 1 also holds for decomposable models. Results on maximum likelihood estimates for log-linear models are usually simplest to state in terms of the ML probabilities $\hat{p}_i$ or corresponding frequencies (or expected counts) $\hat{u}_i = n\hat{p}_i$ rather than the log-linear parameters.

Proposition 2.1.5. *Let $A \in \mathbb{N}^{d \times k}$ and $u \in \mathbb{N}^k$ be a vector of positive counts. The maximum likelihood estimate of the frequencies $\hat{u}$ in the log-linear model $\mathcal{M}_A$ is the unique non-negative solution to the simultaneous system of equations*

$$A\hat{u} = Au \quad \text{and} \quad \hat{u} \in V(I_A).$$

This result was referred to as *Birch's Theorem* in [73, §1.2]. The toric ideal I_A is the lattice ideal $I_{\mathcal{L}}$ in Theorem 1.3.6, and $V(I_A)$ is its affine variety in $\mathbb{R}^k$.

Proof. Let $b_1, \ldots, b_l$ be a basis for $\ker A$. The optimization problem we wish to solve is the constrained optimization problem:

Maximize $u^T \log v$

subject to $b_j^T \log v = 0$ for all $j = 1, \ldots, l$, and $\sum_{i=1}^{k} v_i = n$.

Introducing $l + 1$ Lagrange multipliers $\lambda_1, \ldots, \lambda_l, \gamma$, the critical points are the solutions to the $k + l + 1$ equations

$$\frac{u_i}{v_i} + \sum_{j=1}^{l} \lambda_j \frac{b_{ij}}{v_i} + \gamma = 0, \qquad b_j^T \cdot \log v = 0, \qquad \sum_{i=1}^{k} v_i = n.$$

The last two sets of conditions say that v belongs to the toric variety $V(I_A)$, and that v is a vector of frequencies with sum n. In other words, v belongs to the (rescaled) log-linear model $\mathcal{M}_A$. Upon clearing denominators, the first conditions say that $u + \lambda B = -\gamma v$. Applying A to both sides of this equation implies that $Au = A\hat{u}$. The fact that there is a unique positive solution is a consequence of the fact that the log-likelihood function is strictly concave in $\log(\theta)$ for positive u. Indeed, the log-likelihood function is the difference of a linear function and the logarithm of the partition function, which is easily seen to be strictly convex.

We note that the concavity of the log-likelihood function of a log-linear model is a general property of the class of exponential families that encompasses the discrete case discussed here. See Definition 2.3.11 and [13, 19]. For an alternative proof using entropy maximization we refer to [73, Theorem 1.10]. $\square$

In order to describe the maximum likelihood estimate for a decomposable model, we need the notion of a junction tree.

Definition 2.1.6. Let Γ be a decomposable simplicial complex. A *junction tree* is a tree whose vertices are the facets of Γ, whose edges are labeled by separators in Γ, and such that each edge splits the set of facets of Γ into the two subcomplexes Γ_1, Γ_2 in the decomposition (Γ_1, S, Γ_2).

A junction tree can be obtained by successively using reducible decompositions to break a decomposable complex down to its constituent simplices. However, the junction tree of a decomposable complex is not uniquely determined. For instance, if $\Gamma = [14][24][34]$, two junction trees are $[14]-[24]-[34]$ and $[14]-[34]-[24]$. In both cases all edges correspond to the separator $\{4\}$.

Proposition 2.1.7. *Let Γ be a decomposable simplicial complex. Let u be data such that all marginals along facets of Γ are positive. Let $J(\Gamma)$ be a junction tree for Γ. Then the maximum likelihood estimate of the i-th component of the table of frequencies is given by the formula*

$$\hat{u}_i = \frac{\prod_{F \in V(J(\Gamma))} (u|_F)_{i_F}}{\prod_{S \in E(J(\Gamma))} (u|_S)_{i_S}}.$$

In particular, decomposable models have ML degree 1.

Proof. By Proposition 2.1.5, it suffices to show that the indicated formula for $\hat{u}$ satisfies that constraints $A_\Gamma \hat{u} = A_\Gamma u$, and that (the normalized version of) $\hat{u}$ belongs to the model $\mathcal{M}_\Gamma$. Showing that $A_\Gamma \hat{u} = A_\Gamma u$ is equivalent to showing that $\hat{u}$ and u have the same Γ-marginals. Each marginal can be checked by summing up along the junction tree. The normalized version of $\hat{u}$ belongs to the model because, by grouping each separator with a Γ-marginal that contains it, we deduce the parametrized form for a distribution in $\mathcal{M}_\Gamma$.   □

As an example in which Proposition 2.1.7 applies, consider the 4-chain $\Gamma = [12][23][34]$, which has a unique junction tree. The formula for the maximum likelihood estimates states that

$$\hat{u}_{ijkl} = \frac{u_{ij++} \cdot u_{+jk+} \cdot u_{++kl}}{u_{+j++} \cdot u_{++k+}}.$$

For the complex $\Gamma = [14][24][34]$ mentioned above, we get

$$\hat{u}_{ijkl} = \frac{u_{i++l} \cdot u_{+j+l} \cdot u_{++kl}}{(u_{+++l})^2}$$

regardless of which junction tree is used. In non-decomposable hierarchical log-linear models, however, the ML degree is no longer equal to 1.

Example 2.1.8 (No 3-way interaction). Let $\Gamma = [12][13][23]$ be the simplicial complex for the model of no 3-way interaction, and let $r_1 = r_2 = r_3 = 2$. Using Proposition 2.1.5 and the fact that this model is the degree 4 hypersurface

$$p_{111}p_{122}p_{212}p_{221} - p_{112}p_{121}p_{211}p_{222} = 0,$$

we see that the maximum likelihood estimate of the frequencies $\hat{u}$ can be obtained by solving the following equation in one unknown δ:

$$(u_{111}+\delta)(u_{122}+\delta)(u_{212}+\delta)(u_{221}+\delta) - (u_{112}-\delta)(u_{121}-\delta)(u_{211}-\delta)(u_{222}-\delta) = 0,$$

in which case, $\hat{u} = u + \delta \cdot (1, -1, -1, 1, -1, 1, 1, -1)$. In particular, this shows that the model of no 3-way interaction has ML degree 3 for binary random variables.   □

For more general hierarchical log-linear models there are no closed form expressions in terms of radicals for the maximum likelihood estimates. Equivalently, the maximum likelihood degree is usually more than 5, and the Galois group is usually the full symmetric group (see e.g. [51, Proposition 3]).

Although there is no closed-form formula for maximum likelihood estimates for non-decomposable log-linear models, the log-likelihood function is concave for these models, and any hill-climbing algorithm can be used to compute the ML estimates. One widely-used method is the *iterative proportional scaling* algorithm. We now describe this algorithm for a general log-linear model with matrix $A = (a_1, \ldots, a_k)$ that has all column sums equal to a. We write $\phi \colon \mathbb{R}^d \to \mathbb{R}^k$ for the monomial parametrization whose image is the model $\mathcal{M}_A$. Thus $\phi_i(\theta) = \theta^{a_i}$.

Algorithm 2.1.9 (Iterative proportional scaling).
Input: *The matrix $A \in \mathbb{N}^{d \times k}$, a table of counts $u \in \mathbb{N}^k$, and a tolerance $\epsilon > 0$.*
Output: *Expected counts $\hat{u}$.*
Step 1: *Initialize $v_i = \|u\|_1 \cdot \frac{1}{k}$ for $i = 1, \ldots, k$.*
Step 2: *While $\|Av - Au\|_1 > \epsilon$ do:*

$$\text{For all } i \in [k], \text{ set } v_i := v_i \cdot \left(\frac{\phi_i(Au)}{\phi_i(Av)} \right)^{1/a}.$$

Step 3: *Output $\hat{u} = v$.*

The proof of convergence of iterative proportional scaling is due to Darroch and Ratcliff [27]. For specialized log-linear models, like hierarchical models, it is easy to derive, and show convergence of, variants of the algorithm. These variants depend on rescaling the joint distribution "one marginal at a time."

Gaussian models. Aside from statistical models for discrete random variables, another situation where the algebraic structure of the log-likelihood function arises is in the case of a multivariate normal random vector. An m-dimensional random vector $X \in \mathbb{R}^m$ is distributed according to the *multivariate normal distribution* $\mathcal{N}(\mu, \Sigma)$ if it has the (Lebesgue) density function

$$p_{\mu,\Sigma}(x) = \frac{1}{(2\pi)^{m/2} (\det \Sigma)^{1/2}} \exp \left\{ -\frac{1}{2}(x - \mu)^T \Sigma^{-1}(x - \mu) \right\}, \quad x \in \mathbb{R}^m,$$

where the parameters are a real vector $\mu \in \mathbb{R}^m$ and a symmetric and positive definite matrix Σ. The multivariate normal distribution is sometimes also called the *Gaussian distribution*. The parameter space for the model of all m-dimensional multivariate normal distributions is the set $\mathbb{R}^m \times PD_m$, where PD_m is the cone of $m \times m$ real symmetric positive definite matrices.

The one-dimensional normal distribution is the familiar "bell curve" from Definition 1.1.5, and the m-dimensional normal random vectors are higher dimensional generalizations of this familiar example. The *mean vector* μ determines the center of the distribution and the *covariance matrix* Σ, which defines the elliptical contours of the density function, gives information about the distribution's spread.

A *Gaussian model* is a statistical model comprising multivariate normal distributions. We will typically consider models in the form

$$\mathcal{P}_\Theta = \{ \mathcal{N}(\mu, \Sigma) \ : \ \theta = (\mu, \Sigma) \in \Theta \},$$

where $\Theta \subseteq \mathbb{R}^m \times PD_m$. The *saturated Gaussian model* has $\Theta = \mathbb{R}^m \times PD_m$. Ignoring the normalizing constant, the log-likelihood function for a Gaussian model is

$$\ell_n(\mu, \Sigma) = -\frac{n}{2} \log \det \Sigma - \frac{1}{2} \text{tr} \left(\Sigma^{-1} \cdot \sum_{i=1}^{n} (X^{(i)} - \mu)(X^{(i)} - \mu)^T \right). \tag{2.1.4}$$

The maximum likelihood estimators in the saturated Gaussian model are

$$\hat{\mu} = \bar{X} = \frac{1}{n} \sum_{i=1}^{n} X^{(i)} \quad \text{and} \quad \hat{\Sigma} = S = \frac{1}{n} \sum_{i=1}^{n} (X^{(i)} - \bar{X})(X^{(i)} - \bar{X})^{T}.$$

Here $\bar{X}$ and S are the *sample mean* and *sample covariance matrix*, respectively. The ML estimators can be derived by rewriting the log-likelihood function in (2.1.4) as

$$\ell_n(\mu, \Sigma) = -\frac{n}{2} \log \det \Sigma - \frac{n}{2} \operatorname{tr}(S\Sigma^{-1}) - \frac{n}{2}(\bar{X} - \mu)^T \Sigma^{-1}(\bar{X} - \mu), \qquad (2.1.5)$$

which can be achieved by writing $X^{(i)} - \mu = (X^{(i)} - \bar{X}) - (\bar{X} - \mu)$, multiplying out the products and observing that the n differences $X^{(i)} - \bar{X}$ sum to zero.

We are interested in the algebraic solution of the maximum likelihood estimation problem for submodels $\Theta \subseteq \mathbb{R}^m \times PD_m$. For general subsets of this type, the optimization problem is complicated. However, in the following special situation, the problem reduces to a familiar one. Let Id_m denote the $m \times m$ *identity matrix*.

Proposition 2.1.10. *Suppose that* $\Theta = \Theta_1 \times \{Id_m\}$ *is the parameter space of a Gaussian model. Then the maximum likelihood estimate of the mean parameter* $\hat{\mu}$ *is the point in* Θ_1 *that is closest to the sample mean* $\bar{X}$ *in the* L^2*-norm.*

Proof. When Σ is the identity matrix Id_m, the log-likelihood function reduces to

$$\begin{aligned} \ell_n(\mu) &= -\frac{n}{2} \operatorname{tr} S - \frac{n}{2}(\bar{X} - \mu)^T(\bar{X} - \mu) \\ &= -\frac{n}{2} \operatorname{tr} S - \frac{n}{2} \|\mu - \bar{X}\|_2^2. \end{aligned}$$

Hence, maximizing $\ell_n(\mu)$ over Θ_1 is equivalent to minimizing $\|\mu - \bar{X}\|_2$ over Θ_1. $\quad\square$

Example 2.1.11. Consider the setup of Proposition 2.1.10 and suppose that Θ_1 is a parametric rational curve given by a polynomial parametrization $g : \mathbb{R} \to \mathbb{R}^m$, where the maximal degree of any of the polynomial coordinate functions g_i is d. Then the likelihood equation obtained from the partial derivatives of $\ell_n(\mu)$ is a polynomial equation of degree $2d - 1$:

$$\frac{\partial \ell_n(g(\gamma))}{\partial \gamma} = \sum_{i=1}^{m} (\bar{X}_i - g_i(\gamma)) \frac{\partial g_i}{\partial \gamma}(\mu).$$

In particular, for a generic such map, the model will have ML degree equal to $2d - 1$, and thus is always guaranteed to have a real critical point. $\quad\square$

A situation that occurs frequently is that the Gaussian parameter space has the form $\mathbb{R}^m \times \Theta_2$. In this case, the optimization problem also simplifies:

Proposition 2.1.12. *Suppose that* $\Theta = \mathbb{R}^m \times \Theta_2$. *Then* $\hat{\mu} = \bar{X}$ *and* $\hat{\Sigma}$ *is the maximizer of*

$$\ell_n(\Sigma) = -\frac{n}{2} \log \det \Sigma - \frac{n}{2} \mathrm{tr}(S\Sigma^{-1})$$

in the set Θ_2.

Proof. The inverse Σ^{-1} of the positive definite matrix Σ is also positive definite. Hence $(\bar{X} - \mu)^T \Sigma^{-1}(\bar{X} - \mu) \geq 0$ and equality holds if and only if $\mu = \bar{X}$. $\qquad\square$

The inverse $K = \Sigma^{-1}$ of the covariance matrix Σ is known as the *concentration matrix* or *precision matrix*. Often it is more convenient to use K instead of Σ when parametrizing a Gaussian model. Observing that $\log \det \Sigma = -\log \det K$ we see that the likelihood function $\ell_n(\Sigma)$ becomes the strictly convex function

$$PD_m \to \mathbb{R}, \ K \mapsto \frac{n}{2} \log \det K - \frac{n}{2} \mathrm{tr}\,(SK). \tag{2.1.6}$$

It is thus easier to formulate the algebraic solution to the likelihood equations in $K = \Sigma^{-1}$. We illustrate this in a simple example from the class of undirected Gaussian graphical models, which will be introduced in Chapter 3.

Example 2.1.13. Let $\Theta_2 = \{\Sigma \in PD_4 : (\Sigma^{-1})_{13} = (\Sigma^{-1})_{24} = 0\}$. Then the set $\Theta = \mathbb{R}^m \times \Theta_2$ defines an undirected Gaussian graphical model, namely, the model associated to the cyclic graph with vertex set $V = [4]$ and edges $12, 23, 34, 14$. When parametrizing the model in terms of the concentration matrix $K = \Sigma^{-1}$, the likelihood equations obtained from the partial derivatives of the function in (2.1.6) have the form

$$\frac{1}{\det K} \cdot \frac{\partial}{\partial k_{ij}} \det K - (2 - \delta_{ij})s_{ij} = 0,$$

where δ_{ij} is the Kronecker delta. The following code in **Singular** computes all the complex critical solutions to these likelihood equations for some randomly chosen sample covariance matrix.

```
LIB "solve.lib";
ring R = 0,(k11,k12,k14,k22,k23,k33,k34,k44), dp;
matrix K[4][4] =  k11,k12,0,k14,
                  k12,k22,k23,0,
                  0,k23,k33,k34,
                  k14,0,k34,k44;
intmat X = random(31,4,4);
intmat S = X*transpose(X);
ideal I = jacob(det(K))-det(K)*jacob(trace(K*S));
ideal J = sat(I,det(K))[1];
solve(J);
```

In particular, there are five solutions to the likelihood equations, exactly one of which lies in the positive definite cone. This `Singular` calculation shows that the Gaussian 4-cycle has maximum likelihood degree 5. A conjectured formula for the maximum likelihood degree of the Gaussian m-cycle appears in Problem 7.4.

Note that when defining the ideal I, we *cleared the denominator* $\det K$ from the likelihood equations. Running the command `dim(std(I))` shows that the space of extraneous solutions introduced is a five-dimensional variety in this case. □

The Gaussian model associated with a more general undirected graph prescribes zero entries in the concentration matrix at the non-edges of the graph, and the likelihood equations are of the following form.

Theorem 2.1.14. *Let $G = (V, E)$ be an undirected graph with edge set E. Let*

$$\Theta_2 = \{\Sigma \in PD_m \ : \ (\Sigma^{-1})_{ij} = 0 \ \text{if } ij \notin E\}.$$

The maximum likelihood estimate of Σ, given a positive definite sample covariance matrix S, is the unique positive definite matrix $\hat{\Sigma}$ such that

$$\hat{\Sigma}_{ij} = S_{ij} \ \text{for } ij \in E \ \text{and for } i = j, \quad \text{and} \quad (\hat{\Sigma}^{-1})_{ij} = 0 \ \text{for } ij \notin E.$$

Proof. In terms of the concentration matrix K, the critical equations are

$$\frac{1}{\det K} \cdot \frac{\partial}{\partial k_{ij}} \det K \ = \ (2 - \delta_{ij})s_{ij} \qquad \text{for } ij \in E \ \text{or } i = j.$$

The left-hand side is the cofactor expansion for $(K^{-1})_{ij} = \hat{\Sigma}_{ij}$. □

The above theorem clarifies that maximum likelihood estimation in Gaussian undirected graphical models corresponds to a positive definite matrix completion problem, and the ML degree of the model is the algebraic complexity of this completion problem. Problem 7.4 states a concrete question about this degree.

2.2   Likelihood Equations for Implicit Models

In this section we consider the problem of computing maximum likelihood estimates in an algebraic statistical model for discrete data that is given implicitly by a system of homogeneous polynomial equations. Our exposition assumes some familiarity with projective varieties and their singularities. Some of the background material for this section can be found in the undergraduate textbook by Cox, Little and O'Shea [25].

Let P be a homogeneous prime ideal in the polynomial ring $\mathbb{R}[p_1, \ldots, p_k]$ and $V(P)$ its variety in the complex projective space $\mathbb{P}^{k-1}$. The set of points in $\mathbb{P}^{k-1}$ that have positive real coordinates is identified with the open probability simplex

$$\text{int}(\Delta_{k-1}) \ = \ \{(p_1, \ldots, p_k) \in \mathbb{R}^k \ : \ p_1, \ldots, p_k > 0 \ \text{and } p_1 + \cdots + p_k = 1\}.$$

Our statistical model is the intersection $V_{\text{int}(\Delta)}(P)$ of the projective variety $V(P)$ with the simplex $\text{int}(\Delta_{k-1})$. To avoid degenerate situations, we shall further assume that $V(P)$ is the Zariski closure of $V_{\text{int}(\Delta)}(P)$. This hypothesis is easy to satisfy; for instance, it holds when $V_{\text{int}(\Delta)}(P)$ contains a regular point of $V(P)$.

The data are given by a non-negative integer vector $u = (u_1, \ldots, u_k) \in \mathbb{N}^k$, and we seek to find a model point $p \in V_{\text{int}(\Delta)}(P)$ that maximizes the likelihood of observing these data. Ignoring the multinomial coefficient, this amounts to maximizing the likelihood function

$$L(p) = \frac{p_1^{u_1} \cdots p_k^{u_k}}{(p_1 + \cdots + p_k)^n}, \qquad (2.2.1)$$

where $n = \|u\|_1 = u_1 + \cdots + u_k$ is the sample size. The denominator in (2.2.1) equals 1 on the simplex Δ_{k-1}, so it appears to be redundant. However, we prefer to write $L(p)$ as a ratio of two homogeneous polynomials of the same degree as in (2.2.1). In this form, the likelihood function $L(p)$ becomes a rational function of degree zero, and so is a function on projective space $\mathbb{P}^{k-1}$. This allows us to use projective algebraic geometry [25, §8] to study its restriction to the variety $V(P)$.

Let $V_{\text{reg}}(P)$ denote the set of regular points on the projective variety $V(P)$. This is the complement of the singular locus of $V(P)$. Recall that the singular locus is the set of points such that the Jacobian matrix of a minimal generating set of P fails to have the maximal rank. For a discussion of singularities, with emphasis on the statistical relevance of their tangent cones and their resolutions, we refer the reader to Sections 2.3 and 5.1.

We consider the following open subset of our projective variety $V(P) \subset \mathbb{P}^{k-1}$:

$$\mathcal{U} = V_{\text{reg}}(P) \setminus V\big(p_1 \cdots p_k(p_1 + \cdots + p_k)\big).$$

Definition 2.2.1. The *likelihood locus* Z_u is the set of all points $p \in \mathcal{U}$ such that the gradient $dL(p)$ is zero in the cotangent space of $V(P)$ at p.

Note that the cotangent space of $V(P)$ is the quotient of the k-dimensional vector space spanned by the basis $\{dp_1, \ldots, dp_k\}$ modulo the row space of the Jacobian matrix and the span of the vector $(1, 1, \ldots, 1)$. Thus the likelihood locus Z_u consists of all critical points of the likelihood function (2.2.1). We will show how to compute the ideal of Z_u, using Gröbner bases methods, with the aim of finding the global maximum of $L(p)$ over the model $V_{\text{int}(\Delta)}(P)$.

While most statistical models are parametric in nature, the usefulness of a constrained optimization approach to the likelihood equations comes from the fact that algebraic parametrizations can be implicitized. We will show in two examples how the prime ideal P of a model can be derived from a given parametrization.

Example 2.2.2 (Random censoring revisited). In this example, we explain how to obtain the implicit equation of the random censoring model from Example 2.1.3.

Recall that this model was given by the rational parametrization

$$
\begin{aligned}
p_0 &= \frac{\lambda_s}{\lambda_1 + \lambda_2 + \lambda_s}, \\
p_1 &= \frac{\lambda_1}{\lambda_1 + \lambda_2 + \lambda_s} \cdot \frac{\lambda_s}{\lambda_2 + \lambda_s}, \\
p_2 &= \frac{\lambda_2}{\lambda_1 + \lambda_2 + \lambda_s} \cdot \frac{\lambda_s}{\lambda_1 + \lambda_s}, \\
p_{12} &= \frac{\lambda_1}{\lambda_1 + \lambda_s} \cdot \frac{\lambda_2}{\lambda_2 + \lambda_s} \cdot \frac{\lambda_1 + \lambda_2 + 2\lambda_s}{\lambda_1 + \lambda_2 + \lambda_s}.
\end{aligned}
$$

The model is a parametrized surface in the probability tetrahedron Δ_3. We compute the implicit equation of this surface using the following piece of code for the computer algebra system Macaulay2:

```
S = frac(QQ[t,l1,l2,ls]);
R = QQ[p0,p1,p2,p12];
f = map(S,R,matrix{{
t*ls/(l1+l2+ls),
t*l1*ls/((l1+l2+ls)*(l2+ls)),
t*l2*ls/((l1+l2+ls)*(l1+ls)),
t*l1*l2*(l1+l2+2*ls)/((l1+ls)*(l2+ls)*(l1+l2+ls))}});
P = kernel f
```

The ideal P is a principal ideal. It is generated by one cubic polynomial:

$$
P = \langle 2p_0p_1p_2 + p_1^2p_2 + p_1p_2^2 - p_0^2p_{12} + p_1p_2p_{12} \rangle.
$$

Note that the extra parameter t is used in the parametrization to produce a homogeneous polynomial in the output. Without the parameter t, the extra polynomial $p_0 + p_1 + p_2 + p_{12} - 1$ appears as a generator of P.

In Example 2.1.3 we showed, using the parametrization, that the ML degree of this model is equal to 3. This degree could have been computed, alternatively, using the implicit equation just found and Algorithm 2.2.9 below.   □

Example 2.2.3 (The cheating coin flipper). Here we illustrate the implicit approach to maximum likelihood estimation for a simple mixture model with $k = 5$. In a game of chance, a gambler tosses the same coin four times in a row, and the number of times heads come up are recorded. The possible outcomes are thus 0, 1, 2, 3, or 4. We observe 242 rounds of this game, and we record the outcomes in the data vector $u = (u_0, u_1, u_2, u_3, u_4) \in \mathbb{N}^5$, where u_i is the number of trials with i heads. The sample size is $n = u_0 + u_1 + u_2 + u_3 + u_4 = 242$.

Suppose we suspect that the gambler uses two biased coins, one in each of his sleeves, and he picks one of his two coins with the same probability before each round. We are led to consider the model that is the mixture of a pair of four-times repeated Bernoulli trials. The mixing parameter π is the probability that the gambler picks the coin in his left sleeve. Let α and β be the probability of heads with the left and right coin, respectively. Then the model stipulates that the probabilities of the five outcomes are

$$
\begin{aligned}
p_0 &= \pi(1-\alpha)^4 + (1-\pi)(1-\beta)^4, \\
p_1 &= 4\pi\alpha(1-\alpha)^3 + 4(1-\pi)\beta(1-\beta)^3, \\
p_2 &= 6\pi\alpha^2(1-\alpha)^2 + 6(1-\pi)\beta^2(1-\beta)^2, \\
p_3 &= 4\pi\alpha^3(1-\alpha) + 4(1-\pi)\beta^3(1-\beta), \\
p_4 &= \pi\alpha^4 + (1-\pi)\beta^4.
\end{aligned}
$$

The polynomial p_i represents the probability of seeing i heads in a round. The likelihood of observing the data u in 242 trials equals

$$
\frac{242!}{u_0!u_1!u_2!u_3!u_4!} \cdot p_0^{u_0}p_1^{u_1}p_2^{u_2}p_3^{u_3}p_4^{u_4}. \tag{2.2.2}
$$

Maximum likelihood estimation means maximizing (2.2.2) subject to $0 < \pi, \alpha, \beta < 1$. The likelihood equations for this *unconstrained optimization problem* have infinitely many solutions: there is a line of critical points in the $\alpha = \beta$ plane.

In order to avoid such non-identifiability issues, we replace the given parametric model by its implicit representation. In order to derive this, we introduce the Hankel matrix

$$
Q = \begin{pmatrix} 12p_0 & 3p_1 & 2p_2 \\ 3p_1 & 2p_2 & 3p_3 \\ 2p_2 & 3p_3 & 12p_4 \end{pmatrix},
$$

and we parametrize Q in matrix form as

$$
Q = 12\pi \begin{pmatrix} (1-\alpha)^2 \\ \alpha(1-\alpha) \\ \alpha^2 \end{pmatrix} \begin{pmatrix} (1-\alpha)^2 \\ \alpha(1-\alpha) \\ \alpha^2 \end{pmatrix}^T + 12(1-\pi) \begin{pmatrix} (1-\beta)^2 \\ \beta(1-\beta) \\ \beta^2 \end{pmatrix} \begin{pmatrix} (1-\beta)^2 \\ \beta(1-\beta) \\ \beta^2 \end{pmatrix}^T.
$$

This is a sum of two rank 1 matrices, so we have $\det(Q) = 0$, for all distributions in the model. Since the model is three-dimensional, and $\det(Q)$ is an irreducible polynomial, it follows that the homogeneous prime ideal P is generated by $\det(Q)$.

This analysis shows that we can compute maximum likelihood estimates for this mixture model by solving the following *constrained optimization problem*:

$$
\begin{aligned}
&\text{Maximize } p_0^{u_0}p_1^{u_1}p_2^{u_2}p_3^{u_3}p_4^{u_4} \\
&\text{subject to } \det(Q) = 0 \text{ and } p_0 + \cdots + p_4 = 1.
\end{aligned} \tag{2.2.3}
$$

For an explicit numerical example, suppose that the data vector equals

$$
u = (51, 18, 73, 25, 75).
$$

We compute the maximum likelihood estimates by applying Algorithm 2.2.9 below to the principal ideal $P = \langle \det(Q) \rangle$. The result shows that the ML degree of this model is 12, that is, the exact solution of problem (2.2.3) leads to an algebraic equation of degree 12. The likelihood locus Z_u consists of 12 points over the complex numbers. Six of these critical points are real and have positive coordinates, so they lie in $V_{\mathrm{int}(\Delta)}(P)$. Among the six points, three are local maxima. The global optimum is the point

$$\hat{p} = (\hat{p}_0, \hat{p}_1, \hat{p}_2, \hat{p}_3, \hat{p}_4) = (0.1210412, 0.2566238, 0.2055576, 0.1075761, 0.309201).$$

Using a numerical optimizer in the parameter space (π, α, β) we can check that $\hat{p}$ is indeed in the image of the model parametrization and thus statistically relevant. Note that each coordinate of $\hat{p}$ is an algebraic number of degree 12 over $\mathbb{Q}$. For instance, the first coordinate $0.1210412\ldots$ is a root of the irreducible polynomial

$$
\begin{aligned}
&86380051147019225968509584870284984432\,\hat{p}_0^{12}\\
&-454602785441887444450567415278023475 2\,\hat{p}_0^{11}\\
&-1164799823319409183144588295116554240 0\,\hat{p}_0^{10}\\
&+4436190742596132515699254438021849088\,\hat{p}_0^{9}\\
&+15023570764107338955386437717101 50656\,\hat{p}_0^{8}\\
&-709119707855374420748705486616351616\,\hat{p}_0^{7}\\
&-3010034765409286729565099406988032\,\hat{p}_0^{6}\\
&+305325528901570233400350659177115 2\,\hat{p}_0^{5}\\
&-3827418898307918709555185551156944\,\hat{p}_0^{4}\\
&-1177511359823330514317231682693 6\,\hat{p}_0^{3}\\
&+2948392698982636120937017512598 8\,\hat{p}_0^{2}\\
&-18563781959494070772239445150 45\,\hat{p}_0\\
&+354698720835244808118661478 16 \qquad = \qquad 0.
\end{aligned}
$$

Exact computation of maximum likelihood estimates leads to such equations. □

Returning to our general discussion, suppose that the given prime ideal P is generated by s homogeneous polynomials in the k unknown probabilities:

$$P = \langle g_1, g_2, \ldots, g_s \rangle \subseteq \mathbb{R}[p_1, \ldots, p_k].$$

This homogeneous ideal represents the statistical model $\mathcal{M} = V_{\mathrm{int}(\Delta)}(P)$. We define the *augmented Jacobian matrix* of format $(s+1) \times k$ as follows:

$$
J(p) = \begin{pmatrix}
p_1 & p_2 & \cdots & p_k \\
p_1 \frac{\partial g_1}{\partial p_1} & p_2 \frac{\partial g_1}{\partial p_2} & \cdots & p_k \frac{\partial g_1}{\partial p_k} \\
p_1 \frac{\partial g_2}{\partial p_1} & p_2 \frac{\partial g_2}{\partial p_2} & \cdots & p_k \frac{\partial g_2}{\partial p_k} \\
\vdots & \vdots & \ddots & \vdots \\
p_1 \frac{\partial g_s}{\partial p_1} & p_2 \frac{\partial g_s}{\partial p_2} & \cdots & p_k \frac{\partial g_s}{\partial p_k}
\end{pmatrix}.
$$

Using Lagrange multipliers for our constrained optimization problem, we find:

Proposition 2.2.4. *A point $p \in \mathcal{U}$ is in the likelihood locus Z_u if and only if the data vector u lies in the row span of the augmented Jacobian matrix $J(p)$.*

Proof. Let V_{aff} be the affine subvariety of $\mathbb{C}^k$ defined by the ideal $P + \langle \sum p_i - 1 \rangle$. The Jacobian of V_{aff} is the Jacobian matrix of P augmented by a row of 1s. The likelihood function L has no poles or zeros on $\mathcal{U}$, so the critical points of L are the same as the critical points of $\log(L) = \sum_i u_i \log p_i$ on V_{aff}. A point $p \in \mathcal{U}$ is a critical point of $\log(L)$ if and only if $d\log(L)(p) = (u_1/p_1, \ldots, u_k/p_k)$ is in the row span of the Jacobian of $P + \langle \sum p_i - 1 \rangle$. As $p_i \neq 0$ on the open set $\mathcal{U}$, this condition is equivalent to u being in the row span of $J(p)$. $\qquad\square$

The following corollary can be derived from Proposition 2.2.4.

Corollary 2.2.5. *There is a dense (Zariski) open subset $\mathcal{V} \subset \mathbb{R}^k$ with the property that, for every data vector $u \in \mathcal{V}$, the likelihood locus Z_u consists of finitely many points and the number of these critical points is independent of u. It is called the (implicit) maximum likelihood degree of the model, or of the ideal P.*

The adjective "Zariski" in the corollary implies that $\mathcal{V}$ must contain almost all $u \in \mathbb{N}^k$. The geometric idea behind this corollary is to consider the incidence variety consisting of pairs (p, u) where $p \in Z_u$. This incidence variety is $(k-1)$-dimensional, and it is the projectivization of a vector bundle over $\mathcal{U}$. From this it follows that Z_u is either empty or finite for generic u. See Proposition 3 in [61].

Note that the implicit ML degree in Corollary 2.2.5 can differ from the value of the parametric ML degree from Definition 2.1.4 even for the same model. The implicit ML degree is usually smaller than the parametric ML degree. This discrepancy arises because the implicit ML degree is insensitive to singularities and because the parametrization might not be one-to-one.

We next give a general bound on the ML degree of a model in terms of the degrees of the generators of $P = \langle g_1, g_2, \ldots, g_s \rangle$. We set $d_i = \text{degree}(g_i)$ and

$$D \quad := \quad \sum_{i_1 + \cdots + i_s \leq k - s - 1} d_1^{i_1} d_2^{i_2} \cdots d_s^{i_s}.$$

Here the sum is over all non-negative integer vectors $(i_1, \ldots, i_s)$ whose coordinate sum is at most $k - s - 1$. It is tacitly assumed that $s < k$. If P has k or more generators then we can obtain a bound by replacing P by a suitable subideal. The following result appeared as Theorem 5 in [61].

Theorem 2.2.6. *The ML degree of the model P is bounded above by $Dd_1 d_2 \cdots d_s$. Equality holds when the generators $g_1, g_2, \ldots, g_s$ of P have generic coefficients.*

Example 2.2.7. Suppose that $k = 5$, $s = 1$ and $d_1 = 3$. Then we have $D = 1 + 3 + 3^2 + 3^3 = 40$ and the bound in Theorem 2.2.6 equals $Dd_1 = 120$. Any model that is a cubic hypersurface in Δ_4 has ML degree at most 120. This bound is attained by generic hypersurfaces. However, cubic threefolds arising in statistical applications are very special varieties, and their ML degrees are usually much smaller. For instance, for the coin model in Example 2.2.3 it is 12. $\qquad\square$

Example 2.2.8. Suppose that $s = k - 1$ and the variety $V(P)$ is finite. Theorem 2.2.6 reduces to Bézout's Theorem since $D = 1$ and the bound is $d_1 \cdots d_{k-1}$. □

Proposition 2.2.4 translates into an algorithm for computing the radical ideal I_u of the likelihood locus Z_u. Once the ideal I_u has been constructed, subsequent numerical methods can be used to compute all zeros of I_u and to identify the global maximum of the likelihood function $L(p)$. The following algorithm can be implemented in any computer algebra system that incorporates Gröbner bases.

Algorithm 2.2.9 (Computing the likelihood ideal).
Input: A homogeneous ideal $P \subset \mathbb{R}[p_1, \ldots, p_k]$ and a data vector $u \in \mathbb{N}^k$.
Output: The likelihood ideal I_u for the model $V_{\mathrm{int}(\Delta)}(P)$ and the data u.
Step 1: Compute $c = \mathrm{codim}(P)$ and let Q be the ideal of $(c+1) \times (c+1)$-minors of the Jacobian matrix $(\partial g_i / \partial p_j)$. (The ideal Q defines the singular locus of P.)
Step 2: Compute the kernel of the augmented Jacobian matrix $J(p)$ over the quotient ring $\mathbb{R}[V] = \mathbb{R}[p_1, \ldots, p_k]/P$. This kernel is a submodule M of $\mathbb{R}[V]^k$.
Step 3: Let I'_u be the ideal in $\mathbb{R}[V]$ which is generated by the polynomials $\sum_{i=1}^{k} u_i \phi_i$, where $(\phi_1, \ldots, \phi_r)$ runs over a set of generators for the module M.
Step 4: Compute the likelihood ideal from I'_u by saturation as follows:

$$I_u \quad := \quad \left(I_u : (p_1 \cdots p_n (p_1 + \cdots + p_n) Q)^\infty \right).$$

We refer to [61] for details on the practical implementation of the algorithm, including the delicate work of computing in the quotient ring $\mathbb{R}[V] = \mathbb{R}[p_1 \ldots, p_k]/P$. A small test implementation is shown at the end of this section.

We note that the ML degree of the model $V_{\mathrm{int}(\Delta)}(P)$ is computed by running Algorithm 2.2.9 on randomly chosen data u. The ML degree then equals the degree of the zero-dimensional ideal I_u after Step 4. Here is an example to show this.

Example 2.2.10. We present a small instance of the context-specific independence (CSI) model introduced by Georgi and Schliep in [52]. Let $n = 7$ and consider the following parametrized mixture model for three binary random variables:

$$
\begin{aligned}
p_{111} &= & \pi_1 \, \alpha_1 \beta_1 \gamma_2 &+& \pi_2 \, \alpha_1 \beta_2 \gamma_1 &+& \pi_3 \, \alpha_2 \beta_1 \gamma_1 \\
p_{112} &= & \pi_1 \, \alpha_1 \beta_1 (1 - \gamma_2) &+& \pi_2 \, \alpha_1 \beta_2 (1 - \gamma_1) &+& \pi_3 \, \alpha_2 \beta_1 (1 - \gamma_1) \\
p_{121} &= & \pi_1 \, \alpha_1 (1 - \beta_1) \gamma_2 &+& \pi_2 \, \alpha_1 (1 - \beta_2) \gamma_1 &+& \pi_3 \, \alpha_2 (1 - \beta_1) \gamma_1 \\
p_{122} &= & \pi_1 \, \alpha_1 (1 - \beta_1)(1 - \gamma_2) &+& \pi_2 \, \alpha_1 (1 - \beta_2)(1 - \gamma_1) &+& \pi_3 \, \alpha_2 (1 - \beta_1)(1 - \gamma_1) \\
p_{211} &= & \pi_1 \, (1 - \alpha_1) \beta_1 \gamma_2 &+& \pi_2 \, (1 - \alpha_1) \beta_2 \gamma_1 &+& \pi_3 \, (1 - \alpha_2) \beta_1 \gamma_1 \\
p_{212} &= & \pi_1 \, (1 - \alpha_1) \beta_1 (1 - \gamma_2) &+& \pi_2 \, (1 - \alpha_1) \beta_2 (1 - \gamma_1) &+& \pi_3 \, (1 - \alpha_2) \beta_1 (1 - \gamma_1) \\
p_{221} &= & \pi_1 \, (1 - \alpha_1)(1 - \beta_1) \gamma_2 &+& \pi_2 \, (1 - \alpha_1)(1 - \beta_2) \gamma_1 &+& \pi_3 \, (1 - \alpha_2)(1 - \beta_1) \gamma_1 \\
p_{222} &= \pi_1 \, (1 - \alpha_1)(1 - \beta_1)(1 - \gamma_2) &+& \pi_2 \, (1 - \alpha_1)(1 - \beta_2)(1 - \gamma_1) &+& \pi_3 \, (1 - \alpha_2)(1 - \beta_1)(1 - \gamma_1)
\end{aligned}
$$

where $\pi_3 = (1 - \pi_1 - \pi_2)$. This parametrization is non-identifiable because it has eight parameters but the model is only six-dimensional. It is a hypersurface of degree 4 in the simplex Δ_7. Its defining polynomial is the *hyperdeterminant*

$$
p_{111}^2 p_{222}^2 + p_{121}^2 p_{212}^2 + p_{122}^2 p_{211}^2 + p_{112}^2 p_{221}^2
$$
$$
-2p_{121}p_{122}p_{211}p_{212} - 2p_{112}p_{122}p_{211}p_{221} - 2p_{112}p_{121}p_{212}p_{221} - 2p_{111}p_{122}p_{211}p_{222}
$$
$$
-2p_{111}p_{121}p_{212}p_{222} - 2p_{111}p_{112}p_{221}p_{222} + 4p_{111}p_{122}p_{212}p_{221} + 4p_{112}p_{121}p_{211}p_{222}.
$$

Applying Algorithm 2.2.9 to the principal ideal generated by this polynomial, we find that the hyperdeterminantal model of format $2 \times 2 \times 2$ has ML degree 13. $\square$

A key feature of Algorithm 2.2.9 is that Step 1 and Step 2 are independent of the data u, so they need to be run only once per model. Moreover, these preprocessing steps can be enhanced by applying the saturation of Step 4 already once at the level of the module M, i.e., after Step 2 one can replace M by

$$\tilde{M} \quad := \quad \left(M : (p_1 \cdots p_k (\sum p_i) \cdot Q)^\infty\right) \quad = \quad \mathbb{R}[V]_{g \cdot p_1 \cdots p_k (\sum p_i)} \cdot M \ \cap \ \mathbb{R}[V]^k,$$

for suitable $g \in Q$. Given any particular data vector $u \in \mathbb{N}^k$, one can then use either M or $\tilde{M}$ in Step 3 to define I'_u. The saturation in Step 4 requires some tricks in order to run efficiently. In the numerical experiments in [61], for many models and most data, it sufficed to saturate only once with respect to a single polynomial, as follows:

Step 4': Pick a random $(c + 1) \times (c + 1)$-submatrix of $J(P)$ and let h be its determinant. With some luck, the likelihood ideal I_u will be equal to $(I'_u : h^\infty)$.

Recall that our objective is to compute maximum likelihood estimates.

Algorithm 2.2.11 (Computing the local maxima of the likelihood function).
Input: *The likelihood ideal I_u for the model $V_{\mathrm{int}(\Delta)}(P)$ and the data u.*
Output: *The list of all local maxima of the likelihood function $L(p)$ on $V_{\mathrm{int}(\Delta)}(P)$.*

Step 1: *Assuming that $\dim(I_u) = 0$, compute the solution set Z_u of I_u numerically.*
   For each positive solution $p^* \in Z_u \cap V_{\mathrm{int}(\Delta)}(P)$ perform the following steps:
Step 2: *Solve the linear system $J(p^*)^T \cdot \lambda = u$ to get Lagrange multipliers λ_i^*. The Lagrangian $\mathcal{L} := \log(L(p)) - \sum_{i=1}^k \lambda_i^* g_i(p)$ is a function of p.*
Step 3: *Compute the Hessian $H(p)$ of the Lagrangian $\mathcal{L}(p)$. Compute the restriction of $H(p^*)$ to the tangent space $\mathrm{kernel}(J(p^*))$ of $V(P)$ at the point p^*.*
Step 4: *If the restricted $H(p^*)$ in Step 3 is negative definite, then output p^* with its log-likelihood $\log(L(p^*))$ and the eigenvalues of the restricted $H(p^*)$.*

We close this section by presenting a piece of `Singular` code which implements Algorithm 2.2.9 in its most basic version. This code is not optimized but it can be used to experiment with small models. The model is specified by its ideal P, and the data are specified by a $1 \times k$-matrix u. We here chose the hyperdeterminantal model in Example 2.2.10:

```
LIB "solve.lib";
ring R = 0, (p111,p112,p121,p122,p211,p212,p221,p222), dp;

ideal P = p111^2*p222^2+p121^2*p212^2+p122^2*p211^2+p112^2*p221^2
-2*p121*p122*p211*p212-2*p112*p122*p211*p221-2*p112*p121*p212*p221
-2*p111*p122*p211*p222-2*p111*p121*p212*p222-2*p111*p112*p221*p222
+4*p111*p122*p212*p221+4*p112*p121*p211*p222    ; //  = the model
```

```
matrix u[1][8] = 2,3,5,7,11,13,17,19;                //   = the data
```

```
matrix J = jacob(sum(maxideal(1))+P) * diag(maxideal(1));
matrix I = diag(P, nrows(J));
module M = modulo(J, I);
ideal Iprime = u * M;
int c = nvars(R) - dim(groebner(P));
ideal Isat = sat(Iprime, minor(J, c+1))[1];
ideal IP = Isat, sum(maxideal(1))-1;
solve(IP);
```

The program outputs all 13 critical points of the likelihood function for the data vector $u = (u_{111}, u_{112}, \ldots, u_{222}) = (2, 3, 5, 7, 11, 13, 17, 19)$.

2.3   Likelihood Ratio Tests

Let $X^{(1)}, \ldots, X^{(n)}$ be independent random vectors that all have the same distribution, which is assumed to be unknown but in the statistical model

$$\mathcal{P}_\Theta = \{P_\theta : \theta \in \Theta\}. \tag{2.3.1}$$

The parameter space Θ is assumed to be a subset of $\mathbb{R}^k$. Suppose that using the information provided by the *random sample* $X^{(1)}, \ldots, X^{(n)}$, we wish to test whether or not the true distribution P_θ belongs to some submodel of $\mathcal{P}_\Theta$ that is given by a subset $\Theta_0 \subset \Theta$. Expressed in terms of the parameter θ, we wish to test

$$H_0 : \theta \in \Theta_0 \quad \text{versus} \quad H_1 : \theta \in \Theta \setminus \Theta_0. \tag{2.3.2}$$

In this section we explain the connection between the geometry of Θ_0 and the behavior of a particular approach to testing (2.3.2), namely, the likelihood ratio test from Definition 2.3.1. We assume that the distributions in $\mathcal{P}_\Theta$ all have densities $p_\theta(x)$ with respect to some fixed measure. This ensures that we can define the *log-likelihood function* of the model $\mathcal{P}_\Theta$ in the familiar form

$$\ell_n(\theta) = \sum_{i=1}^n \log p_\theta(X^{(i)}). \tag{2.3.3}$$

Definition 2.3.1. The *likelihood ratio statistic* for the testing problem in (2.3.2) is

$$\lambda_n = 2 \left(\sup_{\theta \in \Theta} \ell_n(\theta) - \sup_{\theta \in \Theta_0} \ell_n(\theta) \right).$$

The *likelihood ratio test* rejects the null hypothesis H_0 if λ_n is "too large."

Since $\Theta_0 \subset \Theta$, the likelihood ratio statistic λ_n is always non-negative. The rationale behind the likelihood ratio test is that large values of λ_n present evidence against H_0 because they indicate that the observed values are much more likely to occur under a distribution P_θ with θ in Θ as opposed to Θ_0.

We remark that exact evaluation of the likelihood ratio statistic λ_n requires the solution of maximum likelihood estimation problems, as discussed in Sections 2.1 and 2.2.

Example 2.3.2. Suppose the model $\mathcal{P}_\Theta$ is the normal distribution family

$$\{\mathcal{N}(\theta, Id_k) \,:\, \theta \in \mathbb{R}^k\},$$

where Id_k is the $k \times k$-identity matrix. The parameter space is $\Theta = \mathbb{R}^k$. The density of $\mathcal{N}(\theta, Id_k)$ is

$$p_\theta(x) \;=\; \frac{1}{\sqrt{(2\pi)^k}} \exp\left\{-\frac{1}{2}\|x - \theta\|_2^2\right\},$$

and, ignoring the normalizing constant, the log-likelihood function of the model is

$$\ell_n(\theta) \;=\; -\frac{1}{2}\sum_{i=1}^n \|X^{(i)} - \theta\|_2^2.$$

Let

$$\bar{X}_n = \frac{1}{n}\sum_{i=1}^n X^{(i)}$$

be the *sample mean*. Then

$$\ell_n(\theta) \;=\; -\frac{n}{2}\|\bar{X}_n - \theta\|_2^2 - \frac{1}{2}\sum_{i=1}^n \|X^{(i)} - \bar{X}_n\|_2^2. \tag{2.3.4}$$

Since we take $\Theta = \mathbb{R}^k$, the likelihood ratio statistic for (2.3.2) is

$$\lambda_n \;=\; n \cdot \inf_{\theta \in \Theta_0} \|\bar{X}_n - \theta\|_2^2. \tag{2.3.5}$$

In other words, λ_n equals the squared Euclidean distance between $\bar{X}_n$ and Θ_0 rescaled by the sample size n. Recall Proposition 2.1.10 and note that $n \cdot \operatorname{tr}(S) = \sum_{i=1}^n \|X^{(i)} - \bar{X}_n\|_2^2$ when comparing to the expressions for $\ell_n(\theta)$ given there. □

In order to turn the likelihood ratio test from Definition 2.3.1 into a practical procedure we need to be able to judge which values of λ_n are too large to be reconcilable with the null hypothesis H_0. As for the chi-square test and Fisher's exact test discussed in Section 1.1, we perform a probability calculation in which we assume that

$$X^{(1)}, X^{(2)}, \ldots, X^{(n)} \sim P_{\theta_0} \tag{2.3.6}$$

are independent and identically distributed according to the distribution indexed by a true parameter $\theta_0 \in \Theta_0$. In this probability calculation we calculate the probability that the (random) likelihood ratio statistic λ_n is as large or larger than the numerical value of the likelihood ratio statistic calculated from some given data set. This probability is referred to as the *p-value* for the likelihood ratio test.

Example 2.3.3. We continue Example 2.3.2 and rewrite the likelihood ratio statistic in (2.3.5) as

$$\lambda_n = \inf_{\theta \in \Theta_0} \| \sqrt{n}(\bar{X}_n - \theta_0) - \sqrt{n}(\theta - \theta_0) \|_2^2. \qquad (2.3.7)$$

This is convenient because if (2.3.6) holds, then $\sqrt{n}(\bar{X}_n - \theta_0)$ is distributed according to $\mathcal{N}(0, Id_k)$. It follows that λ_n has the same distribution as the squared Euclidean distance

$$\inf_{h \in \sqrt{n}(\Theta_0 - \theta_0)} \| Z - h \|_2^2 \qquad (2.3.8)$$

between the translated and rescaled set $\sqrt{n}(\Theta_0 - \theta_0)$ and a random vector $Z \sim \mathcal{N}(0, Id_k)$. Now if Θ_0 is a d-dimensional affine subspace of $\mathbb{R}^k$, then $\sqrt{n}(\Theta_0 - \theta_0)$ is a d-dimensional linear space. It follows from Lemma 2.3.4 below that λ_n has a χ^2-distribution with $\text{codim}(\Theta_0) = k - d$ degrees of freedom (recall Definition 1.1.5). Let λ_{obs} be the numerical value of the likelihood ratio statistic calculated from some given data set. Then, in the affine case, we can compute a p-value for the likelihood ratio test as

$$P_{\theta_0}(\lambda_n \geq \lambda_{\text{obs}}) = P\left(\chi^2_{\text{codim}(\Theta_0)} \geq \lambda_{\text{obs}} \right). \qquad (2.3.9)$$

Lemma 2.3.4. *If Θ_0 is a d-dimensional linear subspace of $\mathbb{R}^k$ and $X \sim \mathcal{N}(0, \Sigma)$ with positive definite covariance matrix Σ, then*

$$\inf_{\theta \in \Theta_0} (X - \theta)^T \Sigma^{-1} (X - \theta)$$

has a χ^2_{k-d}-distribution.

Proof. Using the Cholesky decomposition method, we can find an invertible matrix C such that $C^T C = \Sigma^{-1}$. The affinely transformed random vector $Y = CX$ has a $\mathcal{N}(0, Id_k)$-distribution, and

$$\inf_{\theta \in \Theta_0} (X - \theta)^T \Sigma^{-1} (X - \theta) = \inf_{\gamma \in C\Theta_0} \| Y - \gamma \|_2^2.$$

The linear space $C\Theta_0$ being d-dimensional, there is an orthogonal matrix Q such that $QC\Theta_0 = \mathbb{R}^d \times \{0\}^{k-d}$. Let $Z = QY$. Then

$$\inf_{\gamma \in C\Theta_0} \| Y - \gamma \|_2^2 = \inf_{\gamma \in C\Theta_0} \| Z - Q\gamma \|_2^2 = Z_{k-d+1}^2 + \cdots + Z_k^2. \qquad (2.3.10)$$

Since Z is distributed according to $\mathcal{N}(0, QQ^T) = \mathcal{N}(0, Id_k)$, the sum of squares on the far right in (2.3.10) has a χ^2_{k-d}-distribution. $\qquad \square$

When deriving the chi-square distribution as the distribution of the likelihood ratio statistic in Example 2.3.3, a key argument was that if Θ_0 is an affine space then $\sqrt{n}(\Theta_0 - \theta_0)$ is a fixed linear space for all $\theta_0 \in \Theta_0$ and sample sizes n. Unfortunately, this is no longer true if Θ_0 is not an affine space. However, if the sample size n is large and Θ_0 a smooth manifold, then the distribution of the likelihood ratio statistic can still be approximated by a chi-square distribution.

Example 2.3.5 (Testing a parabola). Consider the normal distribution situation from Examples 2.3.2 and 2.3.3 in the plane, that is, with $k = 2$. Let

$$\Theta_0 = \{\theta = (\theta_1, \theta_2)^T \in \mathbb{R}^2 : \theta_2 = \theta_1^2\}$$

be a parabola as in Figure 2.3.1. If the true parameter $\theta_0 = (\theta_{01}, \theta_{02})^T$ lies on this parabola, that is, if $\theta_{02} = \theta_{01}^2$, then by (2.3.8), the likelihood ratio statistic λ_n is distributed like the squared Euclidean distance between $Z \sim \mathcal{N}(0, Id_2)$ and the translated and rescaled parabola

$$\sqrt{n}(\Theta_0 - \theta_0) = \left\{\theta \in \mathbb{R}^2 : \theta_2 = \frac{\theta_1^2}{\sqrt{n}} + 2\theta_{01}\theta_1\right\}. \tag{2.3.11}$$

The shape of $\sqrt{n}(\Theta_0 - \theta_0)$ depends on both n and θ_0, and thus, so does the distribution of λ_n. This dependence complicates the computation of an analogue of the p-value in (2.3.9). This problem disappears, however, in asymptotic approximations for large sample size n. As n tends to infinity, the set $\sqrt{n}(\Theta_0 - \theta_0)$ in (2.3.11) converges to

$$TC_{\theta_0}(\Theta_0) = \{\theta \in \mathbb{R}^2 : \theta_2 = 2\theta_{01}\theta_1\}. \tag{2.3.12}$$

This is the tangent line of the parabola Θ_0 at θ_0. Convergence of sets means that $TC_{\theta_0}(\Theta_0)$ is the set of accumulation points of sequences (h_n) with $h_n \in \sqrt{n}(\Theta_0 - \theta_0)$ for all n. It can be shown that this entails the convergence in distribution

$$\lambda_n \xrightarrow{D} \inf_{h \in TC_{\theta_0}(\Theta_0)} \|Z - h\|_2^2, \tag{2.3.13}$$

where $Z \sim \mathcal{N}(0, Id_2)$; see [94, Lemma 7.13]. Since $TC_{\theta_0}(\Theta_0)$ is a line, the right-hand side in (2.3.13) has a χ^2-distribution with $\mathrm{codim}(\Theta_0) = 1$ degree of freedom. This provides an asymptotic justification for computing the approximate p-value

$$p = \lim_{n \to \infty} P_{\theta_0}(\lambda_n \geq \lambda_{\mathrm{obs}}) = P\left(\chi^2_{\mathrm{codim}(\Theta_0)} \geq \lambda_{\mathrm{obs}}\right) \tag{2.3.14}$$

when testing the parabola based on a large sample. □

The asymptotics in (2.3.14) are valid more generally when Θ_0 is a smooth manifold in $\mathbb{R}^k$, $k \geq 2$, because only the local geometry of Θ_0, which is captured by linear tangent spaces of fixed dimension, matters for these asymptotics. Moreover, central limit theorems ensure that (2.3.14) remains valid when the underlying model $\mathcal{P}_\Theta$ is not a family of normal distributions with the identity matrix as covariance matrix but some other well-behaved family of distributions.

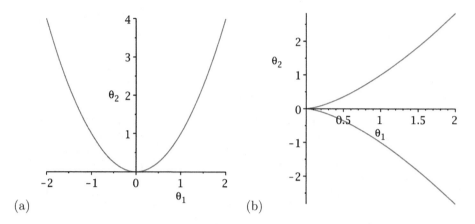

Figure 2.3.1: (a) Parabola and (b) cuspidal cubic.

There are many interesting statistical models whose parameter spaces have singularities. We shall see some explicit examples in Chapter 4 and in Exercise 6.4. See Definition 2.3.15 for the formal definition. It is very important to note that at singularities the limiting distributions of the likelihood ratio statistic need no longer be χ^2-distributions.

Example 2.3.6 (Testing a cuspidal cubic). As in Example 2.3.5, we consider the bivariate normal distributions $\mathcal{N}(\theta, Id_2)$. Suppose that, based on a sample of n observations, we wish to test whether θ lies on the cuspidal cubic

$$\Theta_0 = \{\theta \in \mathbb{R}^2 \ : \ \theta_2^2 = \theta_1^3\}$$

shown on the right-hand side in Figure 2.3.1. At points θ_0 in $\Theta_0 \setminus \{0\}$, the curve Θ_0 has a well-defined tangent line. Therefore, one can show, in exactly the same fashion as for the parabola in Example 2.3.5, that λ_n converges to the χ_1^2-distribution for all true parameters $\theta_0 \in \Theta_0 \setminus \{0\}$.

On the other hand, if the true parameter θ_0 is zero, that is, equal to the sole singularity of Θ_0, then the limiting behavior of λ_n is different. As $n \to \infty$, the sets

$$\sqrt{n}\Theta_0 = \{\theta \in \mathbb{R}^2 \ : \ \theta_2^2 = \theta_1^3/\sqrt{n}\} \tag{2.3.15}$$

converge to the tangent cone

$$TC_0(\Theta_0) = \{\theta \in \mathbb{R}^2 \ : \ \theta_2 = 0, \ \theta_1 \geq 0\}. \tag{2.3.16}$$

From this it is possible to deduce, in analogy to (2.3.13), that λ_n converges in distribution to the squared Euclidean distance between $Z \sim \mathcal{N}(0, Id_2)$ and $TC_0(\Theta_0)$. This squared distance is equal to Z_2^2 if $Z_1 > 0$ and equal to $Z_1^2 + Z_2^2$ if $Z_1 \leq 0$. Since the sign of Z_1 is independent of both Z_2^2 and $Z_1^2 + Z_2^2$, and

$P(Z_1 > 0) = P(Z_1 \leq 0) = 1/2$, it follows that

$$\lambda_n \xrightarrow{D} \frac{1}{2}\chi_1^2 + \frac{1}{2}\chi_2^2, \tag{2.3.17}$$

or in other words,

$$\lim_{n\to\infty} P(\lambda_n \geq \lambda_{\text{obs}}) = \frac{1}{2}P(\chi_1^2 \geq \lambda_{\text{obs}}) + \frac{1}{2}P(\chi_2^2 \geq \lambda_{\text{obs}}) \tag{2.3.18}$$

for all $\lambda_{\text{obs}} > 0$. Consequently, an asymptotic p-value calculated by the chi-square formula (2.3.14) is too small when testing the cuspidal cubic and the true parameter is zero, because $P(\chi_1^2 \geq t) < P(\chi_2^2 \geq t)$ for all t. We remark that singularities can also lead to the chi-square p-value from (2.3.14) being too large. This becomes clear when plotting the curve given by the equation $\theta_2^2 = \theta_1^2 + \theta_1^3$ and considering $\theta_0 = 0$. □

In the examples we discussed above the possible asymptotic distributions for the likelihood ratio statistic were given by distances between a normally distributed random point and tangent lines, or more generally, tangent cones.

Definition 2.3.7. The *tangent cone* $TC_{\theta_0}(\Theta_0)$ of the set $\Theta_0 \subseteq \mathbb{R}^k$ at a point $\theta_0 \in \Theta_0$ is the set of limits of sequences $\alpha_n(\theta_n - \theta_0)$, where α_n are positive reals and $\theta_n \in \Theta$ converge to θ_0. We refer to the elements of a tangent cone as *tangent vectors*.

A tangent cone is a closed set, and it is indeed a cone, that is, multiplying a tangent vector with a non-negative real number yields another tangent vector. Moreover, if $\theta_0 \in \Theta_1 \cap \Theta_2$, then the tangent cone of the union $\Theta_1 \cup \Theta_2$ at θ_0 is the union of the two tangent cones for Θ_1 and Θ_2, respectively.

As we will see in Theorem 2.3.12, distances to tangent cones describe the asymptotic behavior for the likelihood ratio statistic when testing hypotheses described by polynomial equations and inequalities. This leads us to a definition.

Definition 2.3.8. Let $\mathbb{R}[t_1, \ldots, t_k]$ be the ring of polynomials in the indeterminates $t_1, \ldots, t_k$ with real coefficients. A *semi-algebraic set* is a finite union of the form

$$\Theta_0 = \bigcup_{i=1}^{m} \{\theta \in \mathbb{R}^k \mid f(\theta) = 0 \text{ for } f \in F_i \text{ and } h(\theta) > 0 \text{ for } h \in H_i\},$$

where $F_i, H_i \subset \mathbb{R}[t_1, \ldots, t_k]$ are collections of polynomials and all H_i are finite.

Note that all the (sub-)models discussed in this book are given by semi-algebraic sets. Semi-algebraic sets are the basic objects of *real algebraic geometry*. Introductions to real algebraic geometry can be found in the text books [14, 16].

In Examples 2.3.2-2.3.6, we considered the statistical model comprising the normal distributions $\mathcal{N}(\theta, Id_k)$, $\theta \in \mathbb{R}^k$. In this model the behavior of the likelihood ratio statistic is directly connected to the geometry of the null hypothesis defined by a subset $\Theta_0 \subset \mathbb{R}^k$. In more general and possibly non-normal statistical models $\mathcal{P}_\Theta$, the geometry of Θ_0 matters in very similar fashion but in addition we need to take into account how the distributions P_θ (locally) change with θ.

Definition 2.3.9. The *Fisher-information matrix* for the model $\mathcal{P}_\Theta$, $\Theta \subseteq \mathbb{R}^k$, at $\theta \in \Theta$ is the positive semi-definite $k \times k$-matrix $I(\theta)$ with entries

$$I(\theta)_{ij} = \mathrm{E}\left[\left(\frac{\partial}{\partial \theta_i} \log p_\theta(X)\right)\left(\frac{\partial}{\partial \theta_j} \log p_\theta(X)\right)\right], \quad i, j \in [k].$$

The expectation is taken assuming that $X \sim P_\theta$.

In Exercise 6.5 we compute the Fisher-information when θ is the covariance matrix of a centered multivariate normal distribution, and the next example treats the model consisting of all probability distributions on a fixed finite set.

Example 2.3.10 (Discrete Fisher-information). Suppose the sample (2.3.6) consists of discrete random variables taking values in the set $[k+1]$. Let P_θ be the joint distribution of these random variables that is associated with $\theta = (\theta_1, \ldots, \theta_k)$, where θ_i is the probability of observing the value i. Assuming positive distributions, the parameter space is the open probability simplex

$$\Theta = \left\{\theta \in (0,1)^k \ : \ \sum_{i=1}^{k} \theta_i < 1\right\}.$$

The log-density of the distribution P_θ can be expressed as

$$\log p_\theta(x) = \left(\sum_{i=1}^{k} 1_{\{x=i\}} \log \theta_i\right) + 1_{\{x=k+1\}} \log\left(1 - \sum_{i=1}^{k} \theta_i\right).$$

Let $\theta_{k+1} = 1 - \sum_{i=1}^{k} \theta_i$. If $X \sim P_\theta$, then $\mathrm{E}[1_{\{X=i\}}] = \theta_i$. We deduce that the $k \times k$ Fisher-information matrix $I(\theta)$ has i-th diagonal entry equal to $1/\theta_i + 1/\theta_{k+1}$ and all off-diagonal entries equal to $1/\theta_{k+1}$. Its inverse $I(\theta)^{-1}$ is the covariance matrix of the random vector with components $1_{\{X=i\}}$, $i \in [k]$. To check this, we note that $I(\theta)^{-1}$ has diagonal entries $\theta_i(1 - \theta_i)$ and off-diagonal entries $-\theta_i\theta_j$. □

When discussing the behavior of the likelihood ratio statistic some assumptions need to be made about the probabilistic properties of the underlying model $\mathcal{P}_\Theta$. We will assume that $\mathcal{P}_\Theta$ is a regular exponential family, as defined next.

Definition 2.3.11. Let $\mathcal{P}_\Theta = \{P_\theta : \theta \in \Theta\}$ be a family of probability distributions on $\mathcal{X} \subseteq \mathbb{R}^m$ that have densities with respect to a measure ν. We call $\mathcal{P}_\Theta$ an *exponential family* if there is a statistic $T \colon \mathcal{X} \to \mathbb{R}^k$ and functions $h \colon \Theta \to \mathbb{R}^k$ and $Z \colon \Theta \to \mathbb{R}$ such that each distribution P_θ has ν-density

$$p_\theta(x) = \frac{1}{Z(\theta)} \exp\{h(\theta)^T \cdot T(x)\}, \quad x \in \mathcal{X}.$$

Let

$$\Omega = \left\{\omega \in \mathbb{R}^k \ : \ \int_{\mathcal{X}} \exp\{\omega^T T(x)\} \, d\nu(x) < \infty\right\}.$$

If Ω and Θ are open subsets of $\mathbb{R}^k$ and h is a diffeomorphism between Θ and Ω, then we say that $\mathcal{P}_\Theta$ is a *regular exponential family* of order k.

Besides having an open set Θ as parameter space, a regular exponential family enjoys the property that the Fisher-information $I(\theta)$ is well-defined and invertible at all $\theta \in \Theta$. For this and other facts about regular exponential families, we refer the reader to [13, 19]. In this section, regular exponential families simply serve as a device to state a unified result for both multivariate normal and discrete distributions. Indeed the family of all multivariate normal distributions as well as the positive distributions for a discrete random variable discussed in Example 2.3.10 define regular exponential families. How these two examples fall into the framework of Definition 2.3.11 is also explained in detail in [41].

We are now ready to state the main theorem about the asymptotic behavior of the likelihood ratio statistic.

Theorem 2.3.12 (Chernoff). *Suppose the model $\mathcal{P}_\Theta$ is a regular exponential family with parameter space $\Theta \subseteq \mathbb{R}^k$, and let Θ_0 be a semi-algebraic subset of Θ. If the true parameter θ_0 is in Θ_0 and $n \to \infty$, then the likelihood ratio statistic λ_n converges to the distribution of the squared Euclidean distance*

$$\min_{\tau \in TC_{\theta_0}(\Theta_0)} \| Z - I(\theta_0)^{1/2}\tau \|_2^2$$

between the random vector $Z \sim \mathcal{N}(0, Id_k)$ and the linearly transformed tangent cone $I(\theta_0)^{1/2} TC_{\theta_0}(\Theta_0)$. Here $I(\theta_0)^{1/2}$ can be any matrix square root of $I(\theta_0)$.

This theorem has its origins in work by Chernoff [20]. A textbook proof can be found in [94, Theorem 16.7]. The semi-algebraic special case is discussed in [37].

As stated, Chernoff's Theorem covers likelihood ratio tests of a semi-algebraic submodel of a regular exponential family. In the Gaussian case, this amounts to testing a submodel against the saturated model of all multivariate normal distributions, and in the discrete case we test against the model corresponding to the entire probability simplex. However, we may instead be interested in testing

$$H_0: \theta \in \Theta_0 \quad \text{versus} \quad H_1: \theta \in \Theta_1 \setminus \Theta_0 \tag{2.3.19}$$

for two semi-algebraic subsets $\Theta_0 \subset \Theta_1 \subseteq \Theta$, using the likelihood ratio statistic

$$\lambda_n = 2 \left(\sup_{\theta \in \Theta_1} \ell_n(\theta) - \sup_{\theta \in \Theta_0} \ell_n(\theta) \right). \tag{2.3.20}$$

Now the model given by Θ_1 need not be a regular exponential family, and Chernoff's Theorem 2.3.12 does not apply directly. Nevertheless, there is a simple way to determine limiting distributions of the likelihood ratio statistic (2.3.20) when $\Theta_0 \subset \Theta_1 \subseteq \Theta$ and the *ambient model* $\mathcal{P}_\Theta$ is a regular exponential family. We can write the likelihood ratio statistic as the difference of the likelihood ratio statistics for testing (i) Θ_0 versus Θ and (ii) Θ_1 versus Θ. Chernoff's Theorem 2.3.12 now applies to each of the problems (i) and (ii), and we obtain the following corollary.

Corollary 2.3.13 (Testing in a submodel). *Suppose the model $\mathcal{P}_\Theta$ is a regular exponential family with parameter space $\Theta \subseteq \mathbb{R}^k$. Let Θ_0 and Θ_1 be semi-algebraic subsets of Θ. If the true parameter θ_0 is in Θ_0 and $n \to \infty$, then the likelihood ratio statistic λ_n from (2.3.20) converges to the distribution of*

$$\min_{\tau \in TC_{\theta_0}(\Theta_0)} \|Z - I(\theta_0)^{1/2}\tau\|_2^2 - \min_{\tau \in TC_{\theta_0}(\Theta_1)} \|Z - I(\theta_0)^{1/2}\tau\|_2^2,$$

where $Z \sim \mathcal{N}(0, Id_k)$ is a standard normal random vector.

Many statistical models of interest are presented in terms of a parametrization such that $\Theta_0 = g(\Gamma)$ for a map $g \colon \mathbb{R}^d \to \mathbb{R}^k$ and $\Gamma \subseteq \mathbb{R}^d$. If g is a polynomial map and Γ is an open semi-algebraic set, then the Tarski-Seidenberg theorem [14, §2.5.2] ensures that Θ_0 is a semi-algebraic set. In particular, Theorem 2.3.12 applies to such models. Moreover, it is straightforward to compute tangent vectors in $TC_{\theta_0}(\Theta_0)$ directly from the parametrization, by considering the Jacobian matrix

$$J_g(\gamma) = \left(\frac{\partial g_i(\gamma)}{\partial \gamma_j} \right) \in \mathbb{R}^{k \times d}. \tag{2.3.21}$$

Proposition 2.3.14. *If $\theta_0 = g(\gamma_0)$ for some $\gamma_0 \in \Gamma$, then the tangent cone of $\Theta_0 = g(\Gamma)$ at θ_0 contains the linear space spanned by the columns of $J_g(\gamma_0)$.*

Proof. Each vector in the column span of $J_g(\gamma_0)$ is a directional derivative along a curve in $\Theta_0 = g(\Gamma)$, and thus in the tangent cone. $\qquad\square$

When computing the tangent cone $TC_{\theta_0}(\Theta_0)$, it is often useful to complement the given parametrization g with information contained in the implicit representation promised by the Tarski-Seidenberg theorem. In theory, this implicit representation can be computed using algorithms from real algebraic geometry [14], but this tends to be difficult in practice. A subproblem is to find the vanishing ideal

$$\mathcal{I}(\Theta_0) = \{f \in \mathbb{R}[t_1, \ldots, t_k] \; : \; f(\theta) = 0 \text{ for all } \theta \in \Theta_0\}. \tag{2.3.22}$$

Using elimination theory, specifically Gröbner bases and resultants, we can compute a finite generating set $\{f_1, \ldots, f_s\} \subset \mathbb{R}[t_1, \ldots, t_k]$ for the prime ideal $\mathcal{I}(\Theta_0)$; see [25, §3] or [73, §3.2]. From this generating set we can form the Jacobian

$$J_f(\theta) = \left(\frac{\partial f_i}{\partial t_j} \right)_{t=\theta} \in \mathbb{R}^{m \times k}. \tag{2.3.23}$$

Definition 2.3.15. A point θ_0 in $\Theta_0 = g(\Gamma)$ is a *singularity* if the rank of the Jacobian matrix $J_f(\theta_0)$ is smaller than $k - \dim \Theta_0$, the codimension of Θ_0. We note that, in the present setup, $\dim \Theta_0$ equals the rank of $J_g(\gamma)$ for $\gamma \in \Gamma$ generic.

The following lemma describes the simplest case of a tangent cone.

Lemma 2.3.16. *If $\theta_0 = g(\gamma_0)$ is not a singularity of the semi-algebraic set Θ_0 and the rank of the parametric Jacobian $J_g(\gamma_0)$ is equal to $\dim \Theta_0$, then the tangent cone $TC_{\theta_0}(\Theta_0)$ is the linear space spanned by the columns of $J_g(\gamma_0)$.*

At points θ_0 at which Lemma 2.3.16 applies, the limiting distribution of the likelihood ratio statistic in Chernoff's Theorem 2.3.12 is a χ^2-distribution with codim(Θ_0) many degrees of freedom; recall Lemma 2.3.4. Therefore, the asymptotic p-value in (2.3.14) is valid. When considering the setup of Corollary 2.3.13, a χ^2-distribution with $\dim(\Theta_1) - \dim(\Theta_0)$ degrees of freedom arises as a limit when the tangent cones of both Θ_0 and Θ_1 are linear spaces at the true parameter θ_0.

The tangent cone at a singularity can be very complicated. Here, the vanishing ideal $\mathcal{I}(\Theta_0)$ and known polynomial inequalities can be used to find a superset of the tangent cone. Let θ_0 be a root of the polynomial $f \in \mathbb{R}[t_1, \ldots, t_k]$. Write

$$f(t) = \sum_{h=l}^{L} f_h(t - \theta_0)$$

as a sum of homogeneous polynomials f_h in $t - \theta_0 = (t_1 - \theta_{01}, \ldots, t_k - \theta_{0k})$, where $f_h(t)$ has degree h and $f_l \neq 0$. Since $f(\theta_0) = 0$, the minimal degree l is at least 1, and we define $f_{\theta_0,\min} = f_l$.

Lemma 2.3.17. *Suppose θ_0 is a point in the semi-algebraic set Θ_0 and consider a polynomial $f \in \mathbb{R}[t_1, \ldots, t_k]$ such that $f(\theta_0) = 0$ and $f(\theta) \geq 0$ for all $\theta \in \Theta_0$. Then every tangent vector $\tau \in TC_{\theta_0}(\Theta_0)$ satisfies that $f_{\theta_0,\min}(\tau) \geq 0$.*

Proof. Let $\tau \in TC_{\theta_0}(\Theta_0)$ be the limit of the sequence $\alpha_n(\theta_n - \theta_0)$ with $\alpha_n > 0$ and $\theta_n \in \Theta$ converging to θ_0. Let $f_{\theta_0,\min}$ be of degree l. Then the non-negative numbers $\alpha_n^l f(\theta_n)$ are equal to $f_{\theta_0,\min}(\alpha_n(\theta_n - \theta_0))$ plus a term that converges to zero as $n \to \infty$. Thus, $f_{\theta_0,\min}(\tau) = \lim_{n \to \infty} f_{\theta_0,\min}(\alpha_n(\theta_n - \theta_0)) \geq 0$. □

Lemma 2.3.17 applies in particular to every polynomial in the ideal

$$\{f_{\theta_0,\min} \,:\, f \in \mathcal{I}(\Theta_0)\} \subset \mathbb{R}[t_1, \ldots, t_k]. \tag{2.3.24}$$

The algebraic variety defined by this *tangent cone ideal* is the *algebraic tangent cone* of Θ_0, which we denote by $AC_{\theta_0}(\Theta_0)$. Lemma 2.3.17 implies that $TC_{\theta_0}(\Theta_0) \subseteq AC_{\theta_0}(\Theta_0)$. The inclusion is in general strict as can be seen for the cuspidal cubic from Example 2.3.6, where the tangent cone ideal equals $\langle \theta_2^2 \rangle$ and the algebraic tangent cone comprises the entire horizontal axis.

Suppose $\{f^{(1)}, \ldots, f^{(s)}\} \subset \mathbb{R}[t_1, \ldots, t_k]$ is a generating set for the vanishing ideal $\mathcal{I}(\Theta_0)$. Then it is generally not the case that $\{f^{(1)}_{\theta_0,\min}, \ldots, f^{(s)}_{\theta_0,\min}\}$ generates the tangent cone ideal (2.3.24). However, a finite generating set of the ideal in (2.3.24) can be computed using Gröbner basis methods [25, §9.7]. These methods are implemented in `Singular`, which has the command `tangentcone`.

The algebraic techniques just discussed are illustrated in Exercise 6.4, which applies Theorem 2.3.12 to testing a hypothesis about the covariance matrix of a multivariate normal distribution. Other examples can be found in [37]. Many of the models discussed in this book are described by determinantal constraints, and it is an interesting research problem to study their singularities and tangent cones.

Example 2.3.18. Let Θ_0 be the set of positive 3×3-matrices that have rank ≤ 2 and whose entries sum to 1. This semi-algebraic set represents mixtures of two independent ternary random variables; compare Example 4.1.2. The vanishing ideal of Θ_0 equals

$$\mathcal{I}(\Theta_0) \; = \; \left\langle \, t_{11}+t_{12}+t_{13}+t_{21}+t_{22}+t_{23}+t_{31}+t_{32}+t_{33} - 1, \; \det \begin{bmatrix} t_{11} & t_{12} & t_{13} \\ t_{21} & t_{22} & t_{23} \\ t_{31} & t_{32} & t_{33} \end{bmatrix} \right\rangle.$$

The singularities of Θ_0 are precisely those matrices that have rank 1, that is, matrices for which the two ternary random variables are independent (recall Proposition 1.1.2). An example of a singularity is the matrix

$$\theta_0 \; = \; \begin{bmatrix} 1/9 & 1/9 & 1/9 \\ 1/9 & 1/9 & 1/9 \\ 1/9 & 1/9 & 1/9 \end{bmatrix},$$

which determines the uniform distribution. The tangent cone ideal at θ_0 is generated by the sum of the indeterminates

$$t_{11}+t_{12}+t_{13}+t_{21}+t_{22}+t_{23}+t_{31}+t_{32}+t_{33}$$

and the quadric

$$t_{11}t_{22}-t_{12}t_{21}+t_{11}t_{33}-t_{13}t_{31}+t_{22}t_{33}-t_{23}t_{32}-t_{13}t_{22}+t_{12}t_{23}-t_{11}t_{23}+t_{13}t_{21}$$
$$-t_{11}t_{32}+t_{12}t_{31}-t_{22}t_{31}+t_{21}t_{32}-t_{21}t_{33}+t_{23}t_{31}-t_{12}t_{33}+t_{13}t_{32}.$$

We see that the algebraic tangent cone $AC_{\theta_0}(\Theta_0)$ consists of all 3×3-matrices with the property that both the matrix and its adjoint have their respective nine entries sum to zero. Unlike in the case of the cuspidal cubic curve, it can be shown that there are no additional inequalities for the tangent cone at θ_0. Proving this involves an explicit calculation with the multilinear polynomials above. We conclude that

$$AC_{\theta_0}(\Theta_0) \; = \; TC_{\theta_0}(\Theta_0).$$

This equality also holds at any other singularity given by a positive rank 1 matrix θ. However, now the tangent cone comprises all 3×3-matrices whose nine entries sum to zero and whose adjoint $A = (a_{ij})$ satisfies

$$\sum_{i=1}^{3} \sum_{j=1}^{3} a_{ij}\theta_{ji} = 0.$$

This assertion can be verified by running the following piece of **Singular** code:

```
LIB "sing.lib";
ring R = (0,a1,a2,b1,b2),
          (t11,t12,t13, t21,t22,t23, t31,t32,t33),dp;
matrix T[3][3] = t11,t12,t13, t21,t22,t23, t31,t32,t33;
ideal I = det(T),t11+t12+t13+t21+t22+t23+t31+t32+t33-1;
matrix a[3][1] = a1,a2,1-a1-a2;
matrix b[3][1] = b1,b2,1-b1-b2;
matrix t0[3][3] = a*transpose(b);
I = subst(I,t11,t11+t0[1,1],t12,t12+t0[1,2],t13,t13+t0[1,3],
            t21,t21+t0[2,1],t22,t22+t0[2,2],t23,t23+t0[2,3],
            t31,t31+t0[3,1],t32,t32+t0[3,2],t33,t33+t0[3,3]);
                            // shift singularity to origin
tangentcone(I);
```

We invite the reader to extend this analysis to 3×4-matrices and beyond. □

Chapter 3

Conditional Independence

Conditional independence constraints are simple and intuitive restrictions on probability distributions that express the notion that two sets of random variables are unrelated, typically given knowledge of the values of a third set of random variables. A conditional independence model is a family of probability distributions that satisfy a collection of conditional independence constraints. In this chapter we explore the algebraic structure of conditional independence models in the case of discrete or jointly Gaussian random variables. Conditional independence models defined by graphs, known as *graphical models*, are given particular emphasis. Undirected graphical models are also known as *Markov random fields*, whereas directed graphical models are often termed *Bayesian networks*.

This chapter begins with an introduction to general conditional independence models in Section 3.1. We show that in the discrete case and in the Gaussian case conditional independence corresponds to rank constraints on matrices of probabilities and on covariance matrices, respectively. The second and third section both focus on graphical models. Section 3.2 explains the details of how conditional independence constraints are associated with different types of graphs. Section 3.3 describes parametrizations of discrete and Gaussian graphical models. The main results are the Hammersley-Clifford Theorem and the recursive factorization theorem, whose algebraic aspects we explore.

3.1 Conditional Independence Models

Let $X = (X_1, \ldots, X_m)$ be an m-dimensional random vector that takes its values in the Cartesian product $\mathcal{X} = \prod_{j=1}^m \mathcal{X}_j$. We assume throughout that the joint probability distribution of X has a density function $f(x) = f(x_1, \ldots, x_m)$ with respect to a product measure ν on $\mathcal{X}$, and that f is continuous on $\mathcal{X}$. In particular, the continuity assumption becomes ordinary continuity if $\mathcal{X} = \mathbb{R}^m$ and presents no restriction on f if the state space $\mathcal{X}$ is finite. We shall focus on the algebraic

structure of conditional independence in these two settings. For a general introduction to conditional independence (CI) we refer to Milan Studený's monograph
[86].

For each subset $A \subseteq [m]$, let $X_A = (X_a)_{a \in A}$ be the subvector of X indexed
by A. The *marginal density* $f_A(x_A)$ of X_A is obtained by integrating out $x_{[m]\setminus A}$:

$$f_A(x_A) \ := \ \int_{\mathcal{X}_{[m]\setminus A}} f(x_A, x_{[m]\setminus A}) d\nu(x_{[m]\setminus A}), \quad x \in \mathcal{X}_A.$$

Let $A, B \subseteq [m]$ be two disjoint subsets and $x_B \in \mathcal{X}_B$. If $f_B(x_B) > 0$, then the
conditional density of X_A given $X_B = x_B$ is defined as

$$f_{A|B}(x_A|x_B) \ := \ \frac{f_{A \cup B}(x_A, x_B)}{f_B(x_B)}.$$

The conditional density $f_{A|B}(x_A|x_B)$ is undefined when $f_B(x_B) = 0$.

Definition 3.1.1. Let $A, B, C \subseteq [m]$ be pairwise disjoint. The random vector X_A
is *conditionally independent* of X_B given X_C if and only if

$$f_{A \cup B|C}(x_A, x_B|x_C) \ = \ f_{A|C}(x_A|x_C) f_{B|C}(x_B|x_C)$$

for all x_A, x_B and x_C such that $f_C(x_C) > 0$. The notation $X_A \perp\!\!\!\perp X_B \,|\, X_C$ is used
to denote the relationship that X_A is conditionally independent of X_B given X_C.
Often, this is abbreviated to $A \perp\!\!\!\perp B \,|\, C$.

There are a number of immediate consequences of the definition of conditional
independence. These are often called the *conditional independence axioms*.

Proposition 3.1.2. *Let $A, B, C, D \subseteq [m]$ be pairwise disjoint subsets. Then*

(i) *(symmetry)* $X_A \perp\!\!\!\perp X_B \,|\, X_C \implies X_B \perp\!\!\!\perp X_A \,|\, X_C$,

(ii) *(decomposition)* $X_A \perp\!\!\!\perp X_{B \cup D} \,|\, X_C \implies X_A \perp\!\!\!\perp X_B \,|\, X_C$,

(iii) *(weak union)* $X_A \perp\!\!\!\perp X_{B \cup D} \,|\, X_C \implies X_A \perp\!\!\!\perp X_B \,|\, X_{C \cup D}$,

(iv) *(contraction)* $X_A \perp\!\!\!\perp X_B \,|\, X_{C \cup D}$ *and* $X_A \perp\!\!\!\perp X_D \,|\, X_C \implies X_A \perp\!\!\!\perp X_{B \cup D} \,|\, X_C$.

Proof. The proofs of the first three conditional independence axioms (symmetry,
decomposition, and weak union) follow directly from the commutativity of multiplication, marginalization, and conditioning, respectively.

For the proof of the contraction axiom, let x_C be such that $f_C(x_C) > 0$. By
$X_A \perp\!\!\!\perp X_B \,|\, X_{C \cup D}$, we have that

$$f_{A \cup B|C \cup D}(x_A, x_B \,|\, x_C, x_D) \ = \ f_{A|C \cup D}(x_A \,|\, x_C, x_D) \cdot f_{B|C \cup D}(x_B \,|\, x_C, x_D).$$

Multiplying by $f_{C \cup D}(x_C, x_D)$ we deduce that

$$f_{A \cup B \cup C \cup D}(x_A, x_B, x_C, x_D) \ = \ f_{A \cup C \cup D}(x_A, x_C, x_D) \cdot f_{B|C \cup D}(x_B \,|\, x_C, x_D).$$

Dividing by $f_C(x_C) > 0$ we obtain

$$f_{A \cup B \cup D | C}(x_A, x_B, x_D \mid x_C) = f_{A \cup D | C}(x_A, x_D \mid x_C) \cdot f_{B | C \cup D}(x_B \mid x_C, x_D).$$

Applying the conditional independence statement $X_A \perp\!\!\!\perp X_D \mid X_C$, we deduce

$$
\begin{aligned}
f_{A \cup B \cup D | C}(x_A, x_B, x_D \mid x_C) &= f_{A|C}(x_A | x_C) f_{D|C}(x_D | x_C) f_{B|C \cup D}(x_B | x_C, x_D) \\
&= f_{A|C}(x_A \mid x_C) \cdot f_{B \cup D | C}(x_B, x_D \mid x_C),
\end{aligned}
$$

which means that $X_A \perp\!\!\!\perp X_{B \cup D} \mid X_C$. $\qquad\square$

Unlike the first four conditional independence axioms, the following fifth axiom does not hold for every probability density, but only in special cases.

Proposition 3.1.3 (Intersection axiom). *Suppose that $f(x) > 0$ for all x. Then*

$$X_A \perp\!\!\!\perp X_B \mid X_{C \cup D} \text{ and } X_A \perp\!\!\!\perp X_C \mid X_{B \cup D} \implies X_A \perp\!\!\!\perp X_{B \cup C} \mid X_D.$$

Proof. The first and second conditional independence statements imply

$$f_{A \cup B | C \cup D}(x_A, x_B | x_C, x_D) = f_{A | C \cup D}(x_A | x_C, x_D) f_{B | C \cup D}(x_B | x_C, x_D), \quad (3.1.1)$$

$$f_{A \cup C | B \cup D}(x_A, x_C | x_B, x_D) = f_{A | B \cup D}(x_A | x_B, x_D) f_{C | B \cup D}(x_C | x_B, x_D). \quad (3.1.2)$$

Multiplying (3.1.1) by $f_{C \cup D}(x_C, x_D)$ and (3.1.2) by $f_{B \cup D}(x_B, x_D)$, we obtain

$$f_{A \cup B \cup C \cup D}(x_A, x_B, x_C, x_D) = f_{A | C \cup D}(x_A | x_C, x_D) f_{B \cup C \cup D}(x_B, x_C, x_D), \quad (3.1.3)$$

$$f_{A \cup B \cup C \cup D}(x_A, x_B, x_C, x_D) = f_{A | B \cup D}(x_A | x_B, x_D) f_{B \cup C \cup D}(x_B, x_C, x_D). \quad (3.1.4)$$

Equating (3.1.3) and (3.1.4) and dividing by $f_{B \cup C \cup D}(x_B, x_C, x_D)$ (which is allowable since $f(x) > 0$) we deduce that

$$f_{A | C \cup D}(x_A | x_C, x_D) = f_{A | B \cup D}(x_A | x_B, x_D).$$

Since the right-hand side of this expression does not depend on x_C, we conclude

$$f_{A | C \cup D}(x_A | x_C, x_D) = f_{A | D}(x_A | x_D).$$

Plugging this into (3.1.3) and conditioning on X_D gives

$$f_{A \cup B \cup C | D}(x_A, x_B, x_C | x_D) = f_{A | D}(x_A | x_D) f_{B \cup C | D}(x_B, x_C | x_D)$$

and implies that $X_A \perp\!\!\!\perp X_{B \cup C} \mid X_D$. $\qquad\square$

The condition that $f(x) > 0$ for all x is much stronger than necessary for the intersection property to hold. At worst, we only needed to assume that $f_{B \cup C \cup D}(x_B, x_C, x_D) > 0$. However, it is possible to weaken this condition considerably. In the discrete case, it is possible to give a precise characterization of the conditions on the density which guarantee that the intersection property holds. This is described in Exercise 6.6.

Discrete conditional independence models.   Let $X = (X_1, \ldots, X_m)$ be a vector of discrete random variables. Returning to the notation used in previous chapters, we let $[r_j]$ be the set of values taken by X_j. Then X takes its values in $\mathcal{R} = \prod_{j=1}^{m}[r_j]$. In this discrete setting, a conditional independence constraint translates into a system of quadratic polynomial equations in the joint probability distribution.

Proposition 3.1.4. *If X is a discrete random vector, then the conditional independence statement $X_A \!\perp\!\!\!\perp X_B \,|\, X_C$ holds if and only if*

$$p_{i_A,i_B,i_C,+} \cdot p_{j_A,j_B,i_C,+} - p_{i_A,j_B,i_C,+} \cdot p_{j_A,i_B,i_C,+} = 0 \qquad (3.1.5)$$

for all $i_A, j_A \in \mathcal{R}_A$, $i_B, j_B \in \mathcal{R}_B$, and $i_C \in \mathcal{R}_C$.

Proof. By marginalizing we may assume that $A \cup B \cup C = [m]$, and by conditioning we may assume that C is the empty set. By aggregating the states indexed by $\mathcal{R}_A$ and $\mathcal{R}_B$ respectively, we now see that the result follows from Proposition 1.1.2. $\square$

Definition 3.1.5. The *conditional independence ideal* $I_{A \perp\!\!\!\perp B \,|\, C}$ is generated by all quadratic polynomials in (3.1.5).

Equivalently, conditional independence in the discrete case requires each matrix in a certain collection of $\#\mathcal{R}_C$ matrices of size $\#\mathcal{R}_A \times \#\mathcal{R}_B$ to have rank at most 1. The conditional independence ideal $I_{A \perp\!\!\!\perp B \,|\, C}$ is generated by all the 2×2-minors of these matrices. It can be shown that $I_{A \perp\!\!\!\perp B \,|\, C}$ is a prime ideal.

Example 3.1.6 (Marginal independence). The (marginal) independence statement $X_1 \!\perp\!\!\!\perp X_2$, or equivalently, $X_1 \!\perp\!\!\!\perp X_2 \,|\, X_\emptyset$, amounts to saying that the matrix

$$\begin{pmatrix} p_{11} & p_{12} & \cdots & p_{1r_2} \\ p_{21} & p_{22} & \cdots & p_{2r_2} \\ \vdots & \vdots & \ddots & \vdots \\ p_{r_1 1} & p_{r_1 2} & \cdots & p_{r_1 r_2} \end{pmatrix}$$

has rank 1. The independence ideal $I_{1 \perp\!\!\!\perp 2}$ is generated by the 2×2-minors:

$$I_{1 \perp\!\!\!\perp 2} = \big\langle\, p_{i_1 i_2} p_{j_1 j_2} - p_{i_1 j_2} p_{i_2 j_1} \mid i_1, j_1 \in [r_1], i_2, j_2 \in [r_2] \,\big\rangle.$$

For marginal independence, we already saw these quadratic binomial constraints in Chapter 1. $\square$

A *conditional independence model* is the family of distributions that satisfy a set of conditional independence statements $\mathcal{C} = \{A_1 \!\perp\!\!\!\perp B_1 \,|\, C_1,\, A_2 \!\perp\!\!\!\perp B_2 \,|\, C_2, \ldots\}$. Here A_k, B_k, C_k are pairwise disjoint sets for each k. This defines a statistical model in $\Delta_{\mathcal{R}-1}$. The conditional independence ideal of the collection $\mathcal{C}$ is the ideal

$$I_{\mathcal{C}} = I_{A_1 \perp\!\!\!\perp B_1 \,|\, C_1} + I_{A_2 \perp\!\!\!\perp B_2 \,|\, C_2} + \cdots.$$

The conditional independence ideals $I_{\mathcal{C}}$ can be used to investigate implications between conditional independence statements. In particular, one approach to this problem is provided by the primary decomposition of $I_{\mathcal{C}}$.

A *primary decomposition* of an ideal I is a decomposition $I = \cap Q_i$, where each Q_i is a primary ideal and the intersection is irredundant. For the associated algebraic varieties it holds that $V(I) = \cup V(Q_i)$, that is, the variety $V(I)$ is decomposed into its *irreducible components* $V(Q_i)$. The *associated primes* of I are the radicals of the primary ideals Q_i. An associated prime that is minimal with respect to inclusion is called a *minimal prime* of I. The minimal primes are the vanishing ideals of the irreducible components. For more background on primary decomposition see [25, 43]. In the setting of CI models, one hopes that the components of $I_{\mathcal{C}}$ can be understood in terms of conditional independence constraints.

Example 3.1.7 (Conditional and marginal independence). Consider three binary random variables X_1, X_2, and X_3. Consider the collection $\mathcal{C} = \{1 \perp\!\!\!\perp 3 \mid 2, 1 \perp\!\!\!\perp 3\}$. The conditional independence ideal is generated by three quadratic polynomials

$$
\begin{aligned}
I_{\mathcal{C}} &= I_{1 \perp\!\!\!\perp 3 \mid 2} + I_{1 \perp\!\!\!\perp 3} \\
&= \langle p_{111}p_{212} - p_{112}p_{211}, p_{121}p_{222} - p_{122}p_{221}, \\
&\quad (p_{111} + p_{121})(p_{212} + p_{222}) - (p_{112} + p_{122})(p_{211} + p_{221}) \rangle .
\end{aligned}
$$

For binary random variables, these two conditional independence statements are equivalent to saying that the three matrices:

$$
M_1 = \begin{pmatrix} p_{111} & p_{112} \\ p_{211} & p_{212} \end{pmatrix}, \quad M_2 = \begin{pmatrix} p_{121} & p_{122} \\ p_{221} & p_{222} \end{pmatrix}, \quad \text{and} \quad M_1 + M_2 = \begin{pmatrix} p_{1+1} & p_{1+2} \\ p_{2+1} & p_{2+2} \end{pmatrix}
$$

all have rank at most 1. We compute the primary decomposition of $I_{\mathcal{C}}$ in `Singular` with the following code:

```
LIB "primdec.lib";
ring R = 0, (p111,p112,p121,p122,p211,p212,p221,p222), dp;
matrix M1[2][2] = p111,p112,p211,p212;
matrix M2[2][2] = p121,p122,p221,p222;
ideal I = det(M1), det(M2), det(M1 + M2);
primdecGTZ(I);
```

The resulting primary decomposition of $I_{\mathcal{C}}$ can be interpreted in terms of conditional independence constraints:

$$
I_{\mathcal{C}} = I_{1 \perp\!\!\!\perp \{2,3\}} \cap I_{\{1,2\} \perp\!\!\!\perp 3}.
$$

This equation says that, for binary random variables, $1 \perp\!\!\!\perp 3 \mid 2$ and $1 \perp\!\!\!\perp 3$ imply that $1 \perp\!\!\!\perp \{2,3\}$ or $\{1,2\} \perp\!\!\!\perp 3$. A complete exploration of the CI model associated to $\mathcal{C} = \{1 \perp\!\!\!\perp 3 \mid 2, 1 \perp\!\!\!\perp 3\}$ for possibly non-binary variables appears in Exercise 6.7. $\square$

Example 3.1.8 (Failure of the intersection axiom). As alluded to before Proposition 3.1.3, the intersection axiom can fail if the density function is not positive. Here we explore this failure in the case of three binary random variables.

Let X_1, X_2, X_3 be binary random variables, and let $\mathcal{C} = \{1 \perp\!\!\!\perp 2 \mid 3, 1 \perp\!\!\!\perp 3 \mid 2\}$. The conditional independence ideal is generated by four quadratic binomials, which are four of the 2×2-minors of the 2×4 matrix

$$M = \begin{pmatrix} p_{111} & p_{112} & p_{121} & p_{122} \\ p_{211} & p_{212} & p_{221} & p_{222} \end{pmatrix}.$$

The conditional independence ideal $I_\mathcal{C}$ has the primary decomposition:

$$I_\mathcal{C} = I_{1 \perp\!\!\!\perp \{2,3\}} \cap \langle p_{111}, p_{211}, p_{122}, p_{222} \rangle \cap \langle p_{112}, p_{212}, p_{121}, p_{221} \rangle.$$

The first component, $I_{1 \perp\!\!\!\perp \{2,3\}}$, amounts to saying that M is a rank 1 matrix. It is the component which corresponds to the conclusion of the intersection axiom. The other components correspond to families of probability distributions that might not satisfy the conclusion of the intersection axiom. For instance, the second component corresponds to all probability distributions of the form

$$\begin{pmatrix} 0 & p_{112} & p_{121} & 0 \\ 0 & p_{212} & p_{221} & 0 \end{pmatrix}$$

that is, all probability distributions such that $p_{+11} = p_{+22} = 0$.   □

A special class of conditional independence models are the graphical models, described in the next section. These are obtained from a particular collection of conditional independence statements that are derived from combinatorial separation properties in an underlying graph. One reason for preferring these graphical representations is that they often have natural and useful parametrizations, to be discussed in Section 3.3. It is too much to ask that every discrete conditional independence model have a parametrization: independence models need not be irreducible subsets of the probability simplex. However, we might hope that the next best thing holds, as formulated in the following question.

Question 3.1.9. *Is it true that every irreducible component of a conditional independence model has a rational parametrization? In other words, is every irreducible component of a conditional independence model a unirational variety?*

Example 3.1.10. Let X_1, X_2, X_3, X_4 be binary random variables, and consider the conditional independence model

$$\mathcal{C} = \{1 \perp\!\!\!\perp 3 \mid \{2,4\}, 2 \perp\!\!\!\perp 4 \mid \{1,3\}\}.$$

These are the conditional independence statements that hold for the graphical model associated to the four cycle graph with edges $\{12, 23, 34, 14\}$; see Section 3.2. The conditional independence ideal is generated by eight quadratic binomials:

$$\begin{aligned}
I_\mathcal{C} &= I_{1 \perp\!\!\!\perp 3 \mid \{2,4\}} + I_{2 \perp\!\!\!\perp 4 \mid \{1,3\}} \\
&= \langle p_{1111}p_{2121} - p_{1121}p_{2111}, p_{1112}p_{2122} - p_{1122}p_{2112}, \\
&\qquad p_{1211}p_{2221} - p_{1221}p_{2211}, p_{1212}p_{2222} - p_{1222}p_{2212}, \\
&\qquad p_{1111}p_{1212} - p_{1112}p_{1211}, p_{1121}p_{1222} - p_{1122}p_{1221}, \\
&\qquad p_{2111}p_{2212} - p_{2112}p_{2211}, p_{2121}p_{2222} - p_{2122}p_{2221} \rangle.
\end{aligned}$$

The ideal $I_{\mathcal{C}}$ is radical and has nine minimal primes. One of these is a toric ideal I_{Γ}, namely the vanishing ideal of the hierarchical (and graphical) model associated to the simplicial complex $\Gamma = [12][23][34][14]$. The other eight components are linear ideals whose varieties all lie on the boundary of the probability simplex. In particular, all the irreducible components of the variety $V(I_{\mathcal{C}})$ are rational. $\square$

It seems to be a difficult problem in general to address the rationality of conditional independence varieties. One case where an affirmative answer is known is when the CI ideal is a binomial ideal (that is, generated by binomials $p^u - \alpha p^v$). Here rationality holds because of the following result of commutative algebra [44].

Theorem 3.1.11 (Binomial primary decomposition). *Every associated prime of a binomial ideal is a binomial ideal, and the corresponding primary components can be chosen to be binomial ideals as well. In particular, every irreducible component of a binomial variety is a toric variety, and is rational.*

In particular, one can generalize the rationality result of Example 3.1.10 to any model where the given conditional independence statements are all *saturated*:

Corollary 3.1.12. *If $\mathcal{C}$ consists of CI statements of the form $A \perp\!\!\!\perp B \,|\, C$ such that $A \cup B \cup C = [m]$, then every irreducible component of $I_{\mathcal{C}}$ is a rational variety.*

Proof. If $A \cup B \cup C = [m]$ for all $A \perp\!\!\!\perp B \,|\, C \in \mathcal{C}$ then $I_{\mathcal{C}}$ is a binomial ideal. $\square$

Gaussian conditional independence models. It is also natural to ask about conditional independence implications in the case of multivariate normal random vectors. In this case, as well, conditional independence is an algebraic condition.

Proposition 3.1.13. *The conditional independence statement $X_A \perp\!\!\!\perp X_B \,|\, X_C$ holds for a multivariate normal random vector $X \sim \mathcal{N}(\mu, \Sigma)$ if and only if the submatrix $\Sigma_{A \cup C, B \cup C}$ of the covariance matrix Σ has rank $\#C$.*

Proof. If $X \sim \mathcal{N}(\mu, \Sigma)$ follows a multivariate normal distribution, then the conditional distribution of $X_{A \cup B}$ given $X_C = x_C$ is the multivariate normal distribution

$$\mathcal{N}\left(\mu_{A \cup B} + \Sigma_{A \cup B, C}\Sigma_{C,C}^{-1}(x_C - \mu_C), \Sigma_{A \cup B, A \cup B} - \Sigma_{A \cup B, C}\Sigma_{C,C}^{-1}\Sigma_{C, A \cup B}\right).$$

See, for example, [17, §B.6]. The statement $X_A \perp\!\!\!\perp X_B \,|\, X_C$ holds if and only if

$$(\Sigma_{A \cup B, A \cup B} - \Sigma_{A \cup B, C}\Sigma_{C,C}^{-1}\Sigma_{C, A \cup B})_{A,B} = \Sigma_{A,B} - \Sigma_{A,C}\Sigma_{C,C}^{-1}\Sigma_{C,B} = 0.$$

The matrix $\Sigma_{A,B} - \Sigma_{A,C}\Sigma_{C,C}^{-1}\Sigma_{C,B}$ is the Schur complement of the matrix

$$\Sigma_{A \cup C, B \cup C} = \begin{pmatrix} \Sigma_{A,B} & \Sigma_{A,C} \\ \Sigma_{C,B} & \Sigma_{C,C} \end{pmatrix}.$$

Since $\Sigma_{C,C}$ is always invertible (it is positive definite), the Schur complement is zero if and only if the matrix $\Sigma_{A \cup C, B \cup C}$ has rank equal to $\#C$. $\square$

The set of matrices of fixed format with rank $\leq k$ is an irreducible variety, defined by the vanishing of all $(k+1) \times (k+1)$ subdeterminants. In the context of symmetric matrices, the ideal generated by these subdeterminants is a prime ideal [23]. Hence, we obtain nice families of conditional independence ideals.

Definition 3.1.14. Fix pairwise disjoint subsets A, B, C of $[m]$. The *Gaussian conditional independence ideal* $J_{A \perp\!\!\!\perp B \mid C}$ is the following ideal in $\mathbb{R}[\sigma_{ij}, 1 \leq i \leq j \leq m]$:

$$J_{A \perp\!\!\!\perp B \mid C} = \langle\, (\#C + 1) \times (\#C + 1) \text{ minors of } \Sigma_{A \cup C, B \cup C} \,\rangle.$$

Let $\mathcal{C}$ be a collection of conditional independence constraints. The *Gaussian conditional independence model* consists of all jointly normal random variables that satisfy all the conditional independence constraints in $\mathcal{C}$. Each Gaussian conditional independence model corresponds to a semi-algebraic subset of the cone of positive definite matrices PD_m. As in the discrete case, we can explore consequences of conditional independence constraints among Gaussian random variables by looking at the primary decomposition of the conditional independence ideal

$$J_{\mathcal{C}} = J_{A_1 \perp\!\!\!\perp B_1 \mid C_1} + J_{A_2 \perp\!\!\!\perp B_2 \mid C_2} + \cdots$$

associated to the collection $\mathcal{C}$.

Example 3.1.15 (Gaussian conditional and marginal independence). Let $\mathcal{C} = \{1 \perp\!\!\!\perp 3, 1 \perp\!\!\!\perp 3 \mid 2\}$. The conditional independence ideal $J_{\mathcal{C}}$ is generated by two minors:

$$J_{\mathcal{C}} = J_{1 \perp\!\!\!\perp 3 \mid 2} + J_{1 \perp\!\!\!\perp 3} = \langle\, \sigma_{13}, \sigma_{13}\sigma_{22} - \sigma_{12}\sigma_{23} \,\rangle.$$

This ideal has the primary decomposition

$$J_{\mathcal{C}} = \langle\, \sigma_{13}, \sigma_{12}\sigma_{23} \,\rangle = \langle\, \sigma_{12}, \sigma_{13} \,\rangle \cap \langle\, \sigma_{13}, \sigma_{23} \,\rangle = J_{1 \perp\!\!\!\perp \{2,3\}} \cap J_{\{1,2\} \perp\!\!\!\perp 3}.$$

It follows that the implication

$$X_1 \perp\!\!\!\perp X_3 \mid X_2 \text{ and } X_1 \perp\!\!\!\perp X_3 \implies X_1 \perp\!\!\!\perp (X_2, X_3) \text{ or } (X_1, X_2) \perp\!\!\!\perp X_3,$$

holds for multivariate normal random vectors.   □

Example 3.1.16 (Gaussian intersection axiom). Since a multivariate normal random vector has a strictly positive density, the intersection axiom from Proposition 3.1.3 is automatically satisfied. However, the associated CI ideal can have interesting primary components. For example, if $\mathcal{C} = \{1 \perp\!\!\!\perp 2 \mid 3, 1 \perp\!\!\!\perp 3 \mid 2\}$ then

$$\begin{aligned}
J_{\mathcal{C}} &= \langle\, \sigma_{12}\sigma_{33} - \sigma_{13}\sigma_{23}, \sigma_{13}\sigma_{22} - \sigma_{12}\sigma_{23} \,\rangle \\
&= \langle\, \sigma_{12}, \sigma_{13} \,\rangle \cap \left(J_{\mathcal{C}} + \langle\, \sigma_{22}\sigma_{33} - \sigma_{23}^2 \,\rangle \right).
\end{aligned}$$

Note that the extra equation $\sigma_{22}\sigma_{33} - \sigma_{23}^2 = \det(\Sigma_{23,23})$ implies that the set of real symmetric matrices satisfying the equations in the second primary component has empty intersection with the cone of positive definite matrices. It is the first component that corresponds to the conditional independence statement $X_1 \perp\!\!\!\perp (X_2, X_3)$, which is the conclusion of the intersection axiom.

On the other hand, if we were to allow *singular* covariance matrices, then the intersection axiom no longer holds. The second component in the intersection provides examples of singular covariance matrices that satisfy $X_1 \perp\!\!\!\perp X_2 \,|\, X_3$ and $X_1 \perp\!\!\!\perp X_3 \,|\, X_3$ but not $X_1 \perp\!\!\!\perp (X_2, X_3)$. We remark that *singular* covariance matrices correspond to singular multivariate normal distributions. These are concentrated on lower-dimensional affine subspaces of $\mathbb{R}^m$. $\qquad\qquad\square$

Question 3.1.9 about the unirationality of conditional independence models extends also to the Gaussian case. However, aside from binomial ideals, which correspond to conditional independence models where every conditional independence statement $A \perp\!\!\!\perp B \,|\, C$ satisfies $\#C \le 1$, not much is known about this problem.

3.2 Graphical Models

Consider a random vector $X = (X_v \,|\, v \in V)$ together with a simple graph $G = (V, E)$ whose nodes index the components of the random vector. We can then interpret an edge $(v, w) \in E$ as indicating some form of dependence between the random variables X_v and X_w. More precisely, the non-edges of G correspond to conditional independence constraints. These constraints are known as the *Markov properties* of the graph G. The graphical model associated with G is a family of multivariate probability distributions for which these Markov properties hold.

This section gives an overview of three model classes: *undirected graphical models* also known as *Markov random fields*, *directed graphical models* also known as *Bayesian networks*, and *chain graph models*. In each case the graph G is assumed to have no loops, that is, $(v, v) \notin E$ for all $v \in V$, and the differences among models arise from different interpretations given to directed versus undirected edges. Here, an edge $(v, w) \in E$ is *undirected* if $(w, v) \in E$, and it is *directed* if $(w, v) \notin E$. The focus of our discussion will be entirely on conditional independence constraints, and we will make no particular distributional assumption on X but rather refer to the 'axioms' discussed in Section 3.1. The factorization properties of the distributions in graphical models, which also lead to model parametrizations, are the topic of Section 3.3. More background on graphical models can be found in Steffen Lauritzen's book [67] as well as in [24, 26, 42, 99].

Undirected graphs. Suppose all edges in the graph $G = (V, E)$ are undirected. The *undirected pairwise Markov property* associates the following conditional independence constraints with the non-edges of G:

$$X_v \perp\!\!\!\perp X_w \,|\, X_{V \setminus \{v, w\}}, \quad (v, w) \notin E. \qquad (3.2.1)$$

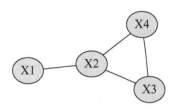

Figure 3.2.1: Undirected graph.

In a multivariate normal distribution $\mathcal{N}(\mu, \Sigma)$ these constraints hold if and only if

$$\det(\Sigma_{(V \setminus \{w\}) \times (V \setminus \{v\})}) = 0 \iff (\Sigma^{-1})_{vw} = 0. \qquad (3.2.2)$$

This equivalence is a special case of Proposition 3.1.13.

The undirected Gaussian graphical model associated with G comprises the distributions $\mathcal{N}(\mu, \Sigma)$ satisfying (3.2.2). For the case when G is a cycle see Example 2.1.13. As we shall see in Proposition 3.3.3, the set of positive joint distributions of discrete random variables that satisfy (3.2.1) coincides with the hierarchical model associated with the simplicial complex whose facets are the maximal cliques of G. Here, a *clique* is any subset of nodes that induces a complete subgraph.

Example 3.2.1. If G is the graph in Figure 3.2.1, then the undirected pairwise Markov property yields the constraints

$$X_1 \perp\!\!\!\perp X_4 \,|\, (X_2, X_3) \quad \text{and} \quad X_1 \perp\!\!\!\perp X_3 \,|\, (X_2, X_4).$$

The multivariate normal distributions in the undirected Gaussian graphical model associated with this graph have concentration matrices $K = \Sigma^{-1}$ with zeros at the $(1, 4)$, $(4, 1)$ and $(1, 3)$, $(3, 1)$ entries. The discrete graphical model is the hierarchical model $\mathcal{M}_\Gamma$ associated with the simplicial complex $\Gamma = [12][234]$. □

The pairwise constraints in (3.2.1) generally entail other conditional independence constraints. These can be determined using the *undirected global Markov property*. This associates with the graph G the constraints $X_A \perp\!\!\!\perp X_B \,|\, X_C$ for all triples of pairwise disjoint subsets $A, B, C \subset V$, A and B non-empty, such that C separates A and B in G. In Example 3.2.1, the global Markov property includes, for instance, the constraint $X_1 \perp\!\!\!\perp (X_3, X_4) \,|\, X_2$.

A joint distribution *obeys* a Markov property if it exhibits the conditional independence constraints that the Markov property associates with the graph.

Theorem 3.2.2 (Undirected global Markov property). *If the random vector X has a joint distribution P^X that satisfies the intersection axiom from Proposition 3.1.3, then P^X obeys the pairwise Markov property for an undirected graph G if and only if it obeys the global Markov property for G.*

The proof of Theorem 3.2.2 given next illustrates the induction arguments that drive many of the results in graphical modelling theory.

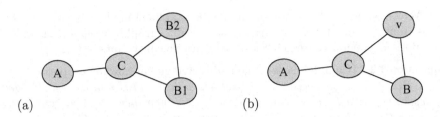

Figure 3.2.2: Illustration of the two cases in the proof of Theorem 3.2.2.

Proof. ($\Longleftarrow$): Any pair of non-adjacent nodes v and w is separated by the complement $V \setminus \{v, w\}$. Hence, the pairwise conditional independence constraints in (3.2.1) are among those listed by the undirected global Markov property.

($\Longrightarrow$): Suppose C separates two non-empty sets A and B. Then the cardinality of $V \setminus C$ is at least 2. If it is equal to 2, then A and B are singletons and $X_A \perp\!\!\!\perp X_B \,|\, X_C$ is one of the pairwise constraints in (3.2.1). This observation provides us with the induction base for an induction on $c = \#(V \setminus C)$. In the induction step $(c - 1) \to c$ we may assume that $c \geq 3$. We distinguish two cases. The high-level structure of the graph in these two cases is depicted in Figure 3.2.2.

Induction step (a): If $A \cup B \cup C = V$, then $c \geq 3$ implies that A or B has at least two elements. Without loss of generality, assume that this is the case for B, which can then be partitioned into two non-empty sets as $B = B_1 \cup B_2$. Then $C \cup B_1$ separates A and B_2. Since the cardinality of $V \setminus (C \cup B_1)$ is smaller than c, the induction assumption implies that

$$X_A \perp\!\!\!\perp X_{B_2} \,|\, (X_C, X_{B_1}). \tag{3.2.3}$$

Swapping the role of B_1 and B_2 we find that

$$X_A \perp\!\!\!\perp X_{B_1} \,|\, (X_C, X_{B_2}). \tag{3.2.4}$$

An application of the intersection axiom to (3.2.3) and (3.2.4) yields the desired constraint $X_A \perp\!\!\!\perp X_B \,|\, X_C$.

Induction step (b): If $A \cup B \cup C \subsetneq V$, then we can choose $v \notin A \cup B \cup C$. In this case $C \cup \{v\}$ separates A and B. By the induction assumption,

$$X_A \perp\!\!\!\perp X_B \,|\, (X_C, X_v). \tag{3.2.5}$$

Any path from A to B intersects C. It follows that $A \cup C$ separates v and B, or $B \cup C$ separates v and A. Without loss of generality, we assume the latter is the case such that the induction assumption implies that

$$X_A \perp\!\!\!\perp X_v \,|\, (X_B, X_C). \tag{3.2.6}$$

The intersection axiom allows us to combine (3.2.5) and (3.2.6) to obtain the constraint $X_A \perp\!\!\!\perp (X_B, X_v) \,|\, X_C$, which implies the desired constraint $X_A \perp\!\!\!\perp X_B \,|\, X_C$.
$\square$

We conclude our discussion of undirected graphical models by showing that for distributions satisfying the intersection axiom, graphical separation indeed determines all general consequences of the pairwise constraints in (3.2.1).

Proposition 3.2.3 (Completeness of the undirected global Markov property). *Suppose $A, B, C \subset V$ are pairwise disjoint subsets with A and B non-empty. If C does not separate A and B in the undirected graph G, then there exists a joint distribution for the random vector X that obeys the undirected global Markov property for G but for which $X_A \perp\!\!\!\perp X_B \mid X_C$ does not hold.*

Proof. We shall prove this statement in the Gaussian case. Consider a path $\pi = (v_1, \ldots, v_n)$ with endpoints $v_1 \in A$ and $v_n \in B$ that does not intersect C. Define a positive definite matrix K by setting all diagonal entries equal to 1, the entries $\{(v_i, v_{i+1}), (v_{i+1}, v_i)\}$ for $i \in [n-1]$ equal to a small non-zero number ρ, and all other entries equal to zero. In other words, the nodes can be ordered such that

$$
K = \begin{pmatrix}
1 & \rho & & & & \\
\rho & 1 & \ddots & & & \\
& \rho & \ddots & \rho & & \\
& & \ddots & 1 & \rho & \\
& & & \rho & 1 & \\
& & & & & Id_{V \setminus \pi}
\end{pmatrix}
$$

where $Id_{V \setminus \pi}$ is the identity matrix of size $\#V - n$.

Let X be a random vector distributed according to $\mathcal{N}(0, \Sigma)$ with $\Sigma = K^{-1}$. By (3.2.2), the distribution of X obeys the pairwise Markov property and thus, by Theorem 3.2.2, also the global Markov property for the graph G. For a contradiction assume that $X_A \perp\!\!\!\perp X_B \mid X_C$. In particular, $X_{v_1} \perp\!\!\!\perp X_{v_n} \mid X_C$. Since $X_{v_1} \perp\!\!\!\perp X_C$, the contraction axiom implies that $X_{v_1} \perp\!\!\!\perp (X_{v_n}, X_C)$. However, this is a contradiction since the absolute value of the cofactor for $\sigma_{v_1 v_n}$ is equal to $|\rho|^{n-1} \neq 0$. $\qquad\square$

Directed acyclic graphs (DAG). Let $G = (V, E)$ be a *directed acyclic graph*, often abbreviated as 'DAG'. The edges are now all directed. The condition of being *acyclic* means that there does not exist a sequence of nodes $v_1, \ldots, v_n$ such that $(v_1, v_2), (v_2, v_3), \ldots, (v_n, v_1)$ are edges in E. The set pa(v) of *parents* of a node $v \in V$ comprises all nodes w such that $(w, v) \in E$. The set de(v) of *descendants* is the set of nodes w such that there is a directed path $(v, u_1), (u_1, u_2), \ldots, (u_n, w)$ from v to w in E. The *non-descendants* of v are nd$(v) = V \setminus (\{v\} \cup \text{de}(v))$.

The *directed local Markov property* associates the CI constraints

$$
X_v \perp\!\!\!\perp X_{\text{nd}(v) \setminus \text{pa}(v)} \mid X_{\text{pa}(v)}, \quad v \in V, \tag{3.2.7}
$$

with the DAG G. The constraints in (3.2.7) reflect the (in-)dependence structure one would expect to observe if the edges represented parent-child or cause-effect relationships; see the two examples in Figure 3.2.3.

Figure 3.2.3: Directed graphs representing (a) $X_1 \!\perp\!\!\!\perp\! X_3 \,|\, X_2$ and (b) $X_1 \!\perp\!\!\!\perp\! X_2$.

What are the general consequences of the conditional independence constraints that the local Markov property associates with a DAG G? As for undirected graphs, this question can be answered by studying separation relations in the graph. However, now a more refined notion of separation is required.

For a subset $C \subseteq V$, we define an(C) to be the set of nodes w that are *ancestors* of some node $v \in C$. Here, w is an ancestor of v if there is a directed path from w to v. In symbols, $v \in$ de(w). Consider an undirected path $\pi = (v_0, v_1, \ldots, v_n)$ in G. This means that, for each i, either (v_i, v_{i+1}) or (v_{i+1}, v_i) is a directed edge of G. If $i \in [n-1]$, then v_i is a non-endpoint node on the path π and we say that v_i is a *collider* on π if the edges incident to v_i are of the form

$$v_{i-1} \longrightarrow v_i \longleftarrow v_{i+1}.$$

For instance, X_3 is a collider on the path from X_1 to X_2 in Figure 3.2.3(b).

Definition 3.2.4. Two nodes v and w in a DAG $G = (V, E)$ are *d-connected* given a conditioning set $C \subseteq V \backslash \{v, w\}$ if there is an undirected path π from v to w such that

(i) all colliders on π are in $C \cup$ an(C), and

(ii) no non-collider on π is in C.

If $A, B, C \subset V$ are pairwise disjoint with A and B non-empty, then C *d-separates* A and B provided no two nodes $v \in A$ and $w \in B$ are d-connected given C.

Example 3.2.5. In the DAG in Figure 3.2.3(a), the singleton $\{X_2\}$ d-separates X_1 and X_3, whereas the empty set d-separates X_1 and X_2 in the DAG in Figure 3.2.3(b). For a little less obvious example, consider the DAG in Figure 3.2.4. In this graph, the nodes X_1 and X_5 are d-separated by $\{X_2\}$, but they are not d-separated by any other subset of $\{X_2, X_3, X_4\}$.   □

We can now define the *directed global Markov property*, which associates with a DAG G the constraints $X_A \!\perp\!\!\!\perp\! X_B \,|\, X_C$ for all triples of pairwise disjoint subsets $A, B, C \subset V$, A and B non-empty, such that C d-separates A and B in G. For this global Markov property, the following analogue to Theorem 3.2.2 holds.

Theorem 3.2.6 (Directed global Markov property). *Any joint distribution P^X for the random vector X obeys the local Markov property for a directed acyclic graph $G = (V, E)$ if and only if it obeys the global Markov property for G.*

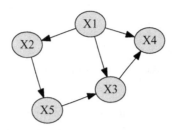

Figure 3.2.4: Directed acyclic graph.

Proof. This result is proven, for example, in [67, §3.2.2]. The proof requires only the contraction axiom from Proposition 3.1.2. As an illustration of how to work with d-separation we present a proof of the easier implication.

($\Longleftarrow$): We need to show that the parent set $\mathrm{pa}(v)$ d-separates a node v from the non-descendants in $\mathrm{nd}(v) \setminus \mathrm{pa}(v)$. For a contradiction, suppose that there is an undirected path $\pi = (v, u_1, \ldots, u_n, w)$ that d-connects v and $w \in \mathrm{nd}(v) \setminus \mathrm{pa}(v)$ given $\mathrm{pa}(v)$. Then $u_1 \notin \mathrm{pa}(v)$ because otherwise it would be a non-collider in the conditioning set $\mathrm{pa}(v)$. Therefore, the edge between v and u_1 points away from v. Since w is a non-descendant of v, there exists a node u_i, $i \in [n]$, that is a collider on π. Let u_j be the collider closest to v, that is, with minimal index j. Since π is d-connecting, u_j is an ancestor of v. This, however, is a contradiction to the acyclicity of G.   □

As a further analogue to the undirected case, we remark that the directed global Markov property is complete: Proposition 3.2.3 remains true if we consider a DAG instead of an undirected graph and replace separation by d-separation.

Finally, we point out a problem that is particular to DAGs. Two distinct DAGs can possess identical d-separation relations and thus encode the exact same conditional independence constraints. The graphs are then termed *Markov equivalent*. For instance, there are two DAGs that are Markov equivalent to the DAG in Figure 3.2.3(a), namely the graphs $X_1 \leftarrow X_2 \rightarrow X_3$ and $X_1 \leftarrow X_2 \leftarrow X_3$. Markov equivalence can be determined efficiently using the following result.

Theorem 3.2.7. *Two directed acyclic graphs $G_1 = (V, E_1)$ and $G_2 = (V, E_2)$ are Markov equivalent if and only if the following two conditions are both met:*

(i) *G_1 and G_2 have the same skeleton, that is, $(v, w) \in E_1 \setminus E_2$ implies $(w, v) \in E_2$ and $(v, w) \in E_2 \setminus E_1$ implies $(w, v) \in E_1$;*

(ii) *G_1 and G_2 have the same unshielded colliders, which are triples of nodes (u, v, w) that induce a subgraph equal to $u \rightarrow v \leftarrow w$.*

A proof of this result can be found in [7, Theorem 2.1], which also addresses the problem of finding a suitable representative of a Markov equivalence class.

Chain graphs. Given the different conditional independence interpretations of undirected and directed graphs, it is natural to ask for a common generalization. Such a generalization is provided by chain graphs, as defined in Definition 3.2.10. However, two distinct conditional independence interpretations of chain graphs have been discussed in the statistical literature. These arise through different specifications of the interplay of directed and undirected edges. The two cases are referred to as LWF or AMP chain graphs in [8], and are called 'block concentrations' and 'concentration regressions' in [98]. Here we will use the two acronyms LWF and AMP, which are the initials of the authors of the original papers: Lauritzen-Wermuth-Frydenberg [49, 68] and Andersson-Madigan-Perlman [8].

In the Gaussian case, the two types of chain graph models always correspond to smooth manifolds in the positive definite cone. In light of Section 2.3, this ensures that chi-square approximations are valid for likelihood ratio tests comparing two chain graph models.

Example 3.2.8 (Gaussian chain graphs). The graph G in Figure 3.2.5(a) is an example of a chain graph. The conditional independence constraints specified by the AMP Markov property for G turn out to be

$$X_1 \perp\!\!\!\perp (X_2, X_4), \qquad X_2 \perp\!\!\!\perp X_4 \mid (X_1, X_3), \qquad (3.2.8)$$

whereas the LWF Markov property yields

$$X_1 \perp\!\!\!\perp (X_2, X_4) \mid X_3, \qquad X_2 \perp\!\!\!\perp X_4 \mid (X_1, X_3). \qquad (3.2.9)$$

We see that, under the LWF interpretation, the chain graph is Markov equivalent to the undirected graph obtained by converting the directed edge between X_1 and X_3 into an undirected edge. Therefore, the associated Gaussian model is the set of multivariate normal distributions whose concentration matrix $K = \Sigma^{-1}$ has zeros over the non-edges of G. The corresponding covariance matrices Σ form a smooth subset of PD_4. The AMP covariance matrices $\Sigma = (\sigma_{ij})$ satisfy

$$\sigma_{12} = \sigma_{14} = 0, \quad \sigma_{13}^2 \sigma_{24} - \sigma_{11}\sigma_{24}\sigma_{33} + \sigma_{11}\sigma_{23}\sigma_{34} = 0.$$

The variety defined by these equations is non-singular over PD_4 because

$$\frac{\partial}{\partial \sigma_{24}} (\sigma_{13}^2 \sigma_{24} - \sigma_{11}\sigma_{24}\sigma_{33} + \sigma_{11}\sigma_{23}\sigma_{34}) = \sigma_{13}^2 - \sigma_{11}\sigma_{33} \neq 0$$

for all $\Sigma = (\sigma_{ij})$ in PD_4. $\square$

By the Markov equivalence between the graph from Figure 3.2.5(a) and the underlying undirected tree, the LWF model for discrete random variables is the hierarchical model $\mathcal{M}_\Gamma$ for $\Gamma = [13][23][34]$. This connection to undirected graphs is more general: the distributions in discrete LWF models are obtained by multiplying together conditional probabilities from several undirected graphical models. In particular, these models are always smooth over the interior of the probability simplex; see the book by Lauritzen [67, §§4.6.1, 5.4.1] for more details.

Discrete AMP models, however, are still largely unexplored and computational algebra provides a way to explore examples and hopefully obtain more general results in the future. A first step in this direction was made in [38]:

Proposition 3.2.9. *If X_1, X_2, X_3, X_4 are binary random variables, then the set of positive joint distributions that obey the AMP Markov property for the graph in Figure 3.2.5(a) is singular exactly at distributions under which X_2, X_4 and the pair (X_1, X_3) are completely independent.*

Before we give the computational proof of this proposition, we comment on its statistical implication. If $\bar{G}$ is the graph obtained by removing the undirected edges from the graph G in Figure 3.2.5(a), then the AMP Markov property for $\bar{G}$ specifies the complete independence of X_2, X_4 and the pair (X_1, X_3). Since this is the singular locus of the binary model associated with G, it follows that with discrete random variables, chi-square approximations can be inappropriate when testing for absence of edges in AMP chain graphs (recall Section 2.3).

Proof of Proposition 3.2.9. The conditional independence relations in (3.2.8) impose rank-one constraints on the table of joint probabilities $p = (p_{i_1 i_2 i_3 i_4}) \in \Delta_{15}$ and the marginal table $p = (p_{i_1 i_2 + i_4}) \in \Delta_7$. Under the assumed positivity, each joint probability factors uniquely as

$$p_{i_1 i_2 i_3 i_4} = p_{i_2 i_3 i_4 | i_1} p_{i_1 +++} := P(X_2 = i_2, X_3 = i_3, X_4 = i_4 \mid X_1 = i_1) P(X_1 = i_1).$$

For $i \in \{1, 2\}$ and a subset $A \subseteq \{2, 3, 4\}$, let

$$q_{A|i} = P(X_j = 1 \text{ for all } j \in A \mid X_1 = i).$$

For each $i \in \{1, 2\}$, the seven probabilities $q_{A|i}$ associated with non-empty sets $A \subseteq \{2, 3, 4\}$ can be used to reparametrize the conditititial distribution of (X_2, X_3, X_4) given $X_1 = i$. We have

$$p_{111|i} = q_{234|i}, \qquad\qquad p_{122|i} = q_{2|i} - q_{23|i} - q_{24|i} + q_{234|i},$$
$$p_{112|i} = q_{23|i} - q_{234|i}, \qquad p_{212|i} = q_{3|i} - q_{23|i} - q_{34|i} + q_{234|i},$$
$$p_{121|i} = q_{24|i} - q_{234|i}, \qquad p_{221|i} = q_{4|i} - q_{24|i} - q_{34|i} + q_{234|i},$$
$$p_{211|i} = q_{34|i} - q_{234|i}, \quad p_{222|i} = 1 - q_{2|i} - q_{3|i} - q_{4|i} + q_{23|i} + q_{24|i} + q_{34|i} - q_{234|i}.$$

This reparametrization is convenient because in the new coordinates the conditional independence $X_1 \perp\!\!\!\perp (X_2, X_4)$ holds in a positive distribution if and only if

$$q_{2|1} = q_{2|2}, \qquad\qquad q_{4|1} = q_{4|2}, \qquad\qquad q_{24|1} = q_{24|2}.$$

We can thus compute with only 11 probabilities, which makes the following calculation of a singular locus in **Singular** feasible. We first load a library and then set up our ring as usual:

```
LIB "sing.lib";
ring R = 0,(q2,q4,q24,q31,q32,q231,q232,q341,q342,q2341,q2342),dp;
```

The second conditional independence constraint $X_2 \perp\!\!\!\perp X_4 \,|\, (X_1, X_3)$ translates into the vanishing of four determinants, and we set up the corresponding ideal:

```
matrix Q11[2][2] = q2341,q231,
                   q341, q31;
matrix Q21[2][2] = q2342,q232,
                   q342, q32;
matrix Q12[2][2] = q24-q2341,q2-q231,
                   q4-q341, 1-q31;
matrix Q22[2][2] = q24-q2342,q2-q232,
                   q4-q342, 1-q32;
ideal I = minor(Q11,2),minor(Q21,2),minor(Q12,2),minor(Q22,2);
```

The next piece of code first computes the singular locus, then saturates to remove components corresponding to distributions on the boundary of the probability simplex, and finally computes the radical ideal:

```
ideal SL = slocus(I);
radical(sat(SL,q31*(1-q31)*q32*(1-q32))[1]);
```

The output of these computations shows that the positive distributions in the singular locus satisfy the equations

$$q_2 q_4 = q_{24}, \qquad q_{3|i} q_2 = q_{23|i}, \qquad q_{3|i} q_4 = q_{34|i}, \qquad q_2 q_{3|i} q_4 = q_{234|i}, \qquad i = 1, 2.$$

These equations determine the complete independence $X_2 \perp\!\!\!\perp X_4 \perp\!\!\!\perp (X_1, X_3)$. $\qquad \square$

We now give the definition of chain graphs and introduce their two Markov properties. Let $G = (V, E)$ be a graph with possibly both directed and undirected edges. Let $(v_0, \ldots, v_n)$ be a sequence of nodes, and define $v_{n+1} = v_0$. This sequence is a *semi-directed cycle* if $(v_i, v_{i+1}) \in E$ for all $0 \le i \le n$, and at least one of the edges is directed, that is, $(v_{i+1}, v_i) \notin E$ for some $0 \le i < n$. For example,

$$v_0 \longrightarrow v_1 \text{ --- } v_2 \longrightarrow v_3 \text{ --- } v_4 \text{ --- } v_0$$

is a semi-directed cycle.

Definition 3.2.10. A graph $G = (V, E)$ with possibly both directed and undirected edges is a *chain graph* if it contains no semi-directed cycles.

Two nodes v and w in a chain graph G are said to be *equivalent* if they are connected by a path composed solely of undirected edges. The equivalence classes of this equivalence relation are known as the *chain components* of G. Let $\mathcal{T}$ be the set of chain components. Then each chain component $T \in \mathcal{T}$ induces a connected undirected subgraph. We define a new graph $\mathcal{D} = (\mathcal{T}, \mathcal{E})$ that has the chain components as nodes, and it has an edge $(T_1, T_2) \in \mathcal{E}$ whenever there exist

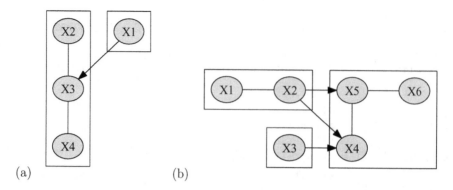

(a)   (b)

Figure 3.2.5: Chain graphs with (a) two and (b) three chain components.

nodes $v_1 \in T_1$ and $v_2 \in T_2$ such that (v_1, v_2) is in the edge set E of the chain graph G. Since G has no semi-directed cycles, the graph $\mathcal{D}$ is a DAG.

Different parent sets will play an important role for the probabilistic interpretation of chain graphs. If T is a chain component and $A \subseteq T$, then we define $\mathrm{pa}_G(A)$ to be the union of all nodes $v \in V \backslash T$ such that $(v, w) \in E$ for some $w \in A$. In contrast, the parent set $\mathrm{pa}_{\mathcal{D}}(T)$ is the union of all chain components $S \in \mathcal{T}$ such that $(S, T) \in \mathcal{E}$. In general, $\mathrm{pa}_G(T) \subsetneq \mathrm{pa}_{\mathcal{D}}(T)$. We write $\mathrm{nd}_{\mathcal{D}}(T)$ for the union of all chain components that are non-descendants of T in $\mathcal{D}$.

Example 3.2.11. The chain graph G in Figure 3.2.5(b) has three chain components enclosed in boxes, namely, $T_1 = \{X_1, X_2\}$, $T_2 = \{X_3\}$ and $T_3 = \{X_4, X_5, X_6\}$. The derived DAG $\mathcal{D}$ is the graph $T_1 \to T_3 \leftarrow T_2$. The parent sets with respect to $\mathcal{D}$ are $\mathrm{pa}_{\mathcal{D}}(T_1) = \mathrm{pa}_{\mathcal{D}}(T_2) = \emptyset$ and $\mathrm{pa}_{\mathcal{D}}(T_3) = \{X_1, X_2, X_3\}$. Note that $\mathrm{pa}_G(T_3) = \{X_2, X_3\}$ is a proper subset of $\mathrm{pa}_{\mathcal{D}}(T_3)$. The non-descendants with respect to $\mathcal{D}$ are $\mathrm{nd}_{\mathcal{D}}(T_1) = \{X_3\}$, $\mathrm{nd}_{\mathcal{D}}(T_2) = \{X_1, X_2\}$ and $\mathrm{nd}_{\mathcal{D}}(T_3) = \{X_1, X_2, X_3\}$. □

The most intuitive versions of the Markov properties for chain graphs are the so-called block-recursive Markov properties. These employ the recursive structure captured by the DAG $\mathcal{D}$. First, they apply a directed Markov property to $\mathcal{D}$. Considering the directed local Markov property from (3.2.7) we obtain the conditional independence constraints

$$X_T \perp\!\!\!\perp X_{\mathrm{nd}_{\mathcal{D}}(T) \backslash \mathrm{pa}_{\mathcal{D}}(T)} \mid X_{\mathrm{pa}_{\mathcal{D}}(T)}, \quad T \in \mathcal{T}. \qquad (3.2.10)$$

Second, for each chain component T, a Markov property for the undirected graph G_T is applied to the conditional distribution of X_T given $X_{\mathrm{pa}_{\mathcal{D}}(T)}$. Using the pairwise Markov property from (3.2.1) we get the constraints

$$X_v \perp\!\!\!\perp X_w \mid (X_{T \backslash \{v,w\}}, X_{\mathrm{pa}_{\mathcal{D}}(T)}), \quad T \in \mathcal{T}, \ v, w \in T. \qquad (3.2.11)$$

Finally, an interpretation is given to the precise structure of the directed edges between chain components. Two non-equivalent interpretations have been considered leading to two different block-recursive Markov properties; see e.g. [8, 98].

Definition 3.2.12. Let $G = (V, E)$ be a chain graph with a set of chain components $\mathcal{T}$ and associated DAG $\mathcal{D} = (\mathcal{T}, \mathcal{E})$. The *AMP block-recursive Markov property* for G specifies the conditional independence constraints (3.2.10), (3.2.11), and

$$X_A \perp\!\!\!\perp X_{\mathrm{pa}_{\mathcal{D}}(T)\backslash \mathrm{pa}_G(A)} \,|\, X_{\mathrm{pa}_G(A)}, \quad T \in \mathcal{T},\ A \subseteq T.$$

The *LWF block-recursive Markov property* for G specifies (3.2.10), (3.2.11), and

$$X_A \perp\!\!\!\perp X_{\mathrm{pa}_{\mathcal{D}}(T)\backslash \mathrm{pa}_G(A)} \,|\, (X_{\mathrm{pa}_G(A)}, X_{\mathrm{nb}(A)}), \quad T \in \mathcal{T},\ A \subseteq T.$$

Here, $\mathrm{nb}(A) = \{v \in T\ :\ (v, w) \in E \text{ for some } w \in A\}$ are the neighbors of A in the undirected graph G_T.

Example 3.2.13. If G is the chain graph G from Figure 3.2.5(b), then (3.2.10) and (3.2.11) each yield precisely one constraint, namely,

$$(X_1, X_2) \perp\!\!\!\perp X_3 \quad \text{and} \quad X_4 \perp\!\!\!\perp X_6 \,|\, (X_1, X_2, X_3, X_5),$$

respectively. The additional constraints specified by the AMP block-recursive Markov property include, for example,

$$(X_5, X_6) \perp\!\!\!\perp (X_1, X_3) \,|\, X_2,$$

which becomes

$$(X_5, X_6) \perp\!\!\!\perp (X_1, X_3) \,|\, (X_2, X_4)$$

in the LWF case. □

Results for AMP and LWF chain graphs include global Markov properties defined using graphical separation criteria, completeness of these global Markov properties, and results on Markov equivalence. Papers that provide entry-points to this topic are [9, 79]. The issue of singularities of discrete AMP chain graph models that we encountered in Proposition 3.2.9 will reappear in Problem 7.10.

3.3 Parametrizations of Graphical Models

Algebraic varieties can be described by polynomials in two different ways, either parametrically or implicitly. For example, the space curve with parametric representation $(x, y, z) = (t^3, t^4, t^5)$ has its implicit representation as a variety $V(P)$ given by the prime ideal $P = \langle y^2 - xz, x^2 y - z^2, x^3 - yz \rangle$. Not every variety has a polynomial parametrization, but many interesting ones do (those that are called *unirational*). As an example of a unirational variety, consider the hyperdeterminantal hypersurface in the space of $2 \times 2 \times 2$-tables, which was parametrized as a context specific independence model in Example 2.2.10. See also Question 3.1.9. The design of algorithms for going back and forth between parametric and implicit representations is an important research area in computational algebra.

The availability of both parametric and implicit representations is also a key feature in the theory of graphical models. For undirected graphical models, the result which makes this relationship precise is the Hammersley-Clifford Theorem. For directed graphical models, the relevant result is the recursive factorization theorem. In Section 3.2, graphical models were introduced via their conditional independence constraints in broad generality. It is possible to give parametric descriptions of graphical models in broad generality. We first present these general descriptions, and then we narrow down to their specific realizations for discrete models and Gaussian models, respectively.

Undirected graphical models.   Let G be an undirected graph on the set of nodes $[m] = \{1, 2, \ldots, m\}$ with edge set E. A *clique* $C \subseteq [m]$ in the graph is a collection of nodes such that $(i, j) \in E$ for every pair $i, j \in C$. The set of maximal cliques is denoted by $\mathcal{C}(G)$. For each $C \in \mathcal{C}(G)$, we introduce a continuous *potential function* $\psi_C(x_C) \geq 0$ which is a function on $\mathcal{X}_C$, the state space of the random vector X_C.

Definition 3.3.1. The *parametrized undirected graphical model* consists of all probability density functions on $\mathcal{X}$ of the form

$$f(x) = \frac{1}{Z} \prod_{C \in \mathcal{C}(G)} \psi_C(x_C) \tag{3.3.1}$$

where

$$Z = \int_{\mathcal{X}} \prod_{C \in \mathcal{C}(G)} \psi_C(x_C) d\nu(x)$$

is the normalizing constant. The parameter space for this model consists of all tuples of potential functions such that the normalizing constant is finite and nonzero. A probability density is said to *factorize* according to the graph G if it can be written in the product form (3.3.1).

The Hammersley-Clifford theorem gives the important result that the parametrized undirected graphical model is the same as the (conditional independence) undirected graphical model from Section 3.2, provided we restrict ourselves to *strictly positive distributions*. For an interesting historical account see Peter Clifford's article [22]; a proof can also be found in [67, Theorem 3.9].

Theorem 3.3.2 (Hammersley-Clifford). *A continuous positive probability density f on $\mathcal{X}$ satisfies the pairwise Markov property on the graph G if and only if it factorizes according to G.*

It is our aim to explore the Hammersley-Clifford theorem from the perspective of algebraic statistics. In particular, we would like to know:

1. What probability distributions come from the factorization/parametrization (3.3.1) in the case of discrete and normal random variables?

2. How can we interpret the Hammersley-Clifford theorem algebraically?

3. Can we use the primary decomposition technique of Section 3.1 to explore the failure of the Hammersley-Clifford theorem for non-negative distributions?

We first focus on the case of discrete random variables $X_1, \ldots, X_m$. Let X_j take its values in $[r_j]$. The joint state space is $\mathcal{R} = \prod_{j=1}^{m} [r_j]$. The graphical model specified by the undirected graph G is a subset of $\Delta_{\mathcal{R}-1}$. In the discrete case, the general parametric description from (3.3.1) becomes a monomial parametrization. Indeed, taking parameters $\theta_{iC}^{C} \in \mathbb{R}_{\geq 0}^{\mathcal{R}_C}$, we have the rational parametrization:

$$p_{i_1 i_2 \cdots i_m} = \phi_{i_1 i_2 \cdots i_m}(\theta) = \frac{1}{Z(\theta)} \prod_{C \in \mathcal{C}(G)} \theta_{iC}^{(C)}. \tag{3.3.2}$$

Proposition 3.3.3. *The parametrized discrete undirected graphical model associated to G consists of all probability distributions in $\Delta_{\mathcal{R}-1}$ of the form $p = \phi(\theta)$ for*

$$\theta = (\theta^{(C)})_{C \in \mathcal{C}(G)} \in \prod_{C \in \mathcal{C}(G)} \mathbb{R}_{\geq 0}^{\mathcal{R}_C}.$$

In particular, the positive part of the parametrized graphical model is precisely the hierarchical log-linear model associated to the simplicial complex of cliques of G.

Denote by I_G the toric ideal of this graphical model. Thus, I_G is the ideal generated by the binomials $p^u - p^v$ corresponding to the Markov basis as in Sections 1.2 and 1.3. We consider the variety $V_\Delta(I_G)$ of the ideal I_G in the closed simplex $\Delta_{\mathcal{R}-1}$. Equivalently, $V_\Delta(I_G)$ consists of all probability distributions on $\mathcal{R}$ that are limits of probability distributions that factor according to the graph. See [51] for a precise polyhedral description of the discrepancy between the set of distributions that factor and the closure of this set. We want to study a coarser problem, namely, comparing $V_\Delta(I_G)$ to conditional independence models $V_\Delta(I_\mathcal{C})$ where $\mathcal{C}$ ranges over conditional independence constraints associated to the graph. Let

$$\text{pairs}(G) = \{i \perp\!\!\!\perp j \,|\, ([m] \setminus \{i,j\}) \;:\; (i,j) \notin E\}$$

be the set of pairwise Markov constraints associated to G and let

$$\text{global}(G) = \{A \perp\!\!\!\perp B \,|\, C \;:\; C \text{ separates } A \text{ from } B \text{ in } G\}$$

be the global Markov constraints associated to G. A graph is called *decomposable* if its complex of cliques is a decomposable simplicial complex (see Definition 1.2.13).

Example 3.3.4. Let G be the graph in Figure 3.2.1. Its Markov properties are

$$\begin{aligned}
\text{pairs}(G) &= \{1 \perp\!\!\!\perp 4 \,|\, \{2,3\}, \, 1 \perp\!\!\!\perp 3 \,|\, \{2,4\}\}, \\
\text{global}(G) &= \text{pairs}(G) \cup \{1 \perp\!\!\!\perp \{3,4\} \,|\, 2\}.
\end{aligned}$$

We consider the case $r_1 = r_2 = r_3 = r_4 = 2$ of four binary random variables. The quadrics described by global(G) are the twelve 2×2-minors of the two matrices

$$M_1 = \begin{pmatrix} p_{1111} & p_{1112} & p_{1122} & p_{1121} \\ p_{2111} & p_{2112} & p_{2122} & p_{2121} \end{pmatrix} \text{ and } M_2 = \begin{pmatrix} p_{1211} & p_{1212} & p_{1222} & p_{1221} \\ p_{2211} & p_{2212} & p_{2222} & p_{2221} \end{pmatrix}.$$

These twelve minors generate a prime ideal of codimension 6. This prime ideal is the conditional independence ideal

$$I_{1 \perp\!\!\!\perp \{3,4\}|2} = \text{minors}(M_1) + \text{minors}(M_2).$$

The maximal cliques C of the graph G are $\{1,2\}$ and $\{2,3,4\}$, so the representation (3.3.2) of this model has $12 = 2^2 + 2^3$ parameters $\theta_{i_C}^{(C)}$. The following code describes the ring map corresponding to ϕ in the algebra software Macaulay2. The partition function Z can be ignored because we are only interested in homogeneous polynomials that belong to the vanishing ideal. We use the following notation for the model parameters: $\theta_{11}^{(12)} = \text{a11}, \theta_{12}^{(12)} = \text{a12}, \ldots, \theta_{221}^{(234)} = \text{b221}, \theta_{222}^{(234)} = \text{b222}$.

```
R = QQ[p1111,p1112,p1121,p1122,p1211,p1212,p1221,p1222,
       p2111,p2112,p2121,p2122,p2211,p2212,p2221,p2222];
S = QQ[a11,a12,a21,a22,b111,b112,b121,b122,b211,b212,b221,b222];
phi = map(S,R, {a11*b111,a11*b112,a11*b121,a11*b122,
               a12*b211,a12*b212,a12*b221,a12*b222,
               a21*b111,a21*b112,a21*b121,a21*b122,
               a22*b211,a22*b212,a22*b221,a22*b222});
P = kernel phi
```

The output of this Macaulay2 code is precisely the prime ideal $I_G = I_{1 \perp\!\!\!\perp \{3,4\}|2}$.

On the other hand, the ideal $I_{\text{pairs}(G)}$ representing the pairwise Markov property on G is not prime. It is properly contained in I_G. Namely, it is generated by eight of the twelve minors of M_1 and M_2, and it can be decomposed as follows:

$$I_{\text{pairs}(G)} = \big(\text{minors}(M_1) \cap \langle p_{1111}, p_{1122}, p_{2111}, p_{2122} \rangle \cap \langle p_{1112}, p_{1121}, p_{2112}, p_{2121} \rangle\big)$$
$$+ \big(\text{minors}(M_2) \cap \langle p_{1211}, p_{1222}, p_{2211}, p_{2222} \rangle \cap \langle p_{1212}, p_{1221}, p_{2212}, p_{2221} \rangle\big).$$

It follows that $I_{\text{pairs}(G)}$ is the intersection of nine prime ideals. One of these primes is $I_G = I_{\text{global}(G)}$. Each of the other eight primes contains one of the unknowns p_{ijkl} which means its variety lies on the boundary of the probability simplex Δ_{15}.

This computation confirms Theorem 3.3.2, and it shows how the conclusion of Proposition 3.3.3 fails on the boundary of Δ_{15}. There are eight such "failure components", one for each associated prime of $I_{\text{pairs}(G)}$. For instance, the prime

$$\langle p_{1111}, p_{1122}, p_{2111}, p_{2122} \rangle + \langle p_{1211}, p_{1222}, p_{2211}, p_{2222} \rangle$$

represents the family of all distributions such that $P(X_3 = X_4) = 0$. All such probability distributions satisfy the pairwise Markov constraints on G but they are not in the closure of the image of the parametrization ϕ.   □

In general, even throwing in all the polynomials implied by global(G) might not be enough to characterize the probability distributions that are limits of factoring distributions. Indeed, this failure occurred for the four-cycle graph in Example 3.1.10. For decomposable graphs, however, everything works out nicely [51].

Theorem 3.3.5. *The following conditions on an undirected graph G are equivalent:*

(i) $I_G = I_{\text{global}(G)}$.

(ii) I_G *is generated by quadrics.*

(iii) *The ML degree of $V(I_G)$ is 1.*

(iv) *G is a decomposable graph.*

Let us now take a look at Gaussian undirected graphical models. The density of the multivariate normal distribution $\mathcal{N}(\mu, \Sigma)$ can be written as

$$f(x) = \frac{1}{Z} \prod_{i=1}^{m} \exp\left\{-\frac{1}{2}(x_i - \mu_i)^2 k_{ii}\right\} \prod_{1 \leq i < j \leq m} \exp\left\{-\frac{1}{2}(x_i - \mu_i)(x_j - \mu_j)k_{ij}\right\},$$

where $K = (k_{ij}) = \Sigma^{-1}$ is the concentration matrix, and Z is a normalizing constant. In particular, we see that the density f always factorizes into pairwise potentials, and that f factorizes as in (3.3.1) if and only if $k_{ij} = 0$ for all $(i, j) \notin E$. This leads to the following observation.

Proposition 3.3.6. *The parametrized Gaussian undirected graphical model corresponds to the set of pairs $(\mu, \Sigma) \in \mathbb{R}^m \times PD_m$ with $(\Sigma^{-1})_{ij} = 0$ for all $(i, j) \notin E$.*

From the linear parametrization of the concentration matrices in the model, we can employ the classical adjoint formula for the inverse to deduce a rational parametrization of the covariance matrices that belong to the model.

Example 3.3.7. Let T be the tree with $V = [4]$ and $E = \{(1, 2), (1, 3), (1, 4)\}$. The concentration matrices in the Gaussian undirected graphical model for T are the positive definite matrices of the form

$$K = \begin{pmatrix} k_{11} & k_{12} & k_{13} & k_{14} \\ k_{12} & k_{22} & 0 & 0 \\ k_{13} & 0 & k_{33} & 0 \\ k_{14} & 0 & 0 & k_{44} \end{pmatrix}.$$

Applying the adjoint formula, we find that the corresponding covariance matrices Σ have the form

$$\Sigma = \frac{1}{\det K} \begin{pmatrix} k_{22}k_{33}k_{44} & -k_{12}k_{33}k_{44} & -k_{13}k_{22}k_{44} & -k_{14}k_{22}k_{33} \\ -k_{12}k_{33}k_{44} & k_{11}k_{33}k_{44} & -k_{12}k_{13}k_{44} & -k_{12}k_{14}k_{33} \\ -k_{13}k_{22}k_{44} & -k_{12}k_{13}k_{44} & k_{11}k_{22}k_{44} & -k_{13}k_{14}k_{22} \\ -k_{14}k_{22}k_{33} & -k_{12}k_{14}k_{33} & -k_{13}k_{14}k_{22} & k_{11}k_{22}k_{33} \end{pmatrix}.$$

It follows that the set of covariance matrices in this Gaussian undirected tree model determine a toric variety (since it is given by a monomial parametrization). Direct computation in this case shows that the vanishing ideal J_T of this set of covariance matrices is equal to the ideal $J_{\mathrm{global}(T)}$ which is generated by quadrics.   □

It is unknown whether the characterization of decomposable graphs on discrete random variables given in Theorem 3.3.5 can be extended in a meaningful way to Gaussian undirected graphical models. For directed graphs see (3.3.7).

Directed graphical models.   In the remainder of this section we consider a directed acyclic graph G on the vertex set $[m]$ and discuss parametrizations of the associated directed graphical models. For each node j of G, we introduce a conditional distribution $f_j(x_j \mid x_{\mathrm{pa}(j)})$ and consider probability densities of the form

$$f(x) = \prod_{j=1}^{m} f_j(x_j \mid x_{\mathrm{pa}(j)}). \tag{3.3.3}$$

Definition 3.3.8. The *parametric directed graphical model* consists of all probability densities that factorize as the product of conditionals (3.3.3).

Here the situation is even better than in the undirected case: factorizations are equivalent to satisfying the local or global Markov property [67, Thm. 3.27].

Theorem 3.3.9 (Recursive factorization). *A probability density f satisfies the recursive factorization property* (3.3.3) *with respect to the directed acyclic graph G if and only if it satisfies the local Markov property.*

In the discrete case, the parametric representation of the directed graphical model G takes the form

$$\phi : \; p_{i_1 i_2 \cdots i_m} \; = \; \prod_{j=1}^{m} \theta^{(j)}(i_j \mid i_{\mathrm{pa}(j)}), \tag{3.3.4}$$

where the parameters $\theta^{(j)}(i_j \mid i_{\mathrm{pa}(j)})$ represent conditional probabilities. They are assumed to satisfy the following linear constraints for all tuples $i_{\mathrm{pa}(j)}$ in $\mathcal{R}_{\mathrm{pa}(j)}$:

$$\sum_{k=1}^{r_j} \theta^{(j)}(k \mid i_{\mathrm{pa}(j)}) \; = \; 1.$$

We write $\phi_{\geq 0}$ for the restriction of the map ϕ to the cone of non-negative parameters or, more precisely, to the product of simplices given by these linear constraints. For instance, in Example 3.3.11 the parameter space is the 9-dimensional cube Δ_1^9.

The *local Markov property* associated with the directed acyclic graph G is the set of conditional independence statements seen in (3.2.7). We abbreviate

$$\mathrm{local}(G) \;\; = \;\; \{\, u \perp\!\!\!\perp \big(\mathrm{nd}(u) \setminus \mathrm{pa}(u)\big) \mid \mathrm{pa}(u) \; : \; u = 1, 2, \ldots, n \,\}. \tag{3.3.5}$$

In the discrete case, these CI statements translate into a system of homogeneous quadratic polynomials. As in Definition 3.1.5, we consider their ideal $I_{\text{local}(G)}$, and we write $V_\Delta(I_{\text{local}(G)})$ for the corresponding variety in the closed simplex $\Delta_{\mathcal{R}-1}$. The discrete recursive factorization theorem takes the following form.

Theorem 3.3.10. *The image of the parametrization $\phi_{\geq 0}$ equals the set of all discrete probability distributions which satisfy the local Markov property. In symbols,*

$$\text{image}(\phi_{\geq 0}) = V_\Delta(I_{\text{local}(G)}).$$

In Theorem 3.3.10 we may replace the local Markov property, $\text{local}(G)$, by the global Markov property, $\text{global}(G)$. The latter was defined in Section 3.2. However, the given formulation is a priori stronger since $\text{local}(G) \subseteq \text{global}(G)$. The next example illustrates the statement of Theorem 3.3.10.

Example 3.3.11. Let G be the DAG depicted in Figure 3.3.1, where the nodes are binary random variables. The local Markov property for this directed graph equals

$$\text{local}(G) = \{ 2 \perp\!\!\!\perp 3 \mid 1, \ 4 \perp\!\!\!\perp 1 \mid \{2,3\} \}.$$

The ideal generated by the quadrics associated with these two conditional independence statements is

$$
\begin{aligned}
I_{\text{local}(G)} = \big\langle \ & (p_{1111} + p_{1112})(p_{1221} + p_{1222}) - (p_{1121} + p_{1122})(p_{1211} + p_{1212}), \\
& (p_{2111} + p_{2112})(p_{2221} + p_{2222}) - (p_{2121} + p_{2122})(p_{2211} + p_{2212}), \\
& p_{1111}p_{2112} - p_{1112}p_{2111}, \ p_{1121}p_{2122} - p_{1122}p_{2121}, \\
& p_{1211}p_{2212} - p_{1212}p_{2211}, \ p_{1221}p_{2222} - p_{1222}p_{2221} \big\rangle.
\end{aligned}
$$

The ideal $I_{\text{local}(G)}$ is prime, and its projective variety is irreducible and has dimension 9. In particular, the implicitly defined model $V_\Delta(I_{\text{local}(G)})$ has no "failure components" in the boundary of the probability simplex. We have the equality

$$I_{\text{local}(G)} = I_{\text{global}(G)} = I_G.$$

The recursive factorization theorem expresses this directed graphical model as the image of a polynomial map $\phi_{\geq 0}$ into the simplex Δ_{15}. We abbreviate the vector of $9 = 2^0 + 2^1 + 2^1 + 2^2$ parameters for this model as follows:

$$\theta = (a, b_1, b_2, c_1, c_2, d_{11}, d_{12}, d_{21}, d_{22}).$$

The letters a, b, c, d correspond to the random variables X_1, X_2, X_3, X_4 in this order. The parameters represent the conditional probabilities of each node given its parents. For instance, the parameter d_{21} was denoted $\theta^{(4)}(1|21)$ in (3.3.4), and it represents the conditional probability of the event "$X_4 = 1$" given "$X_2 = 2$"

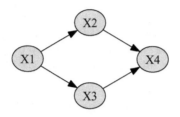

Figure 3.3.1: Directed acyclic graph.

and $X_3 = 1$". With this notation, the coordinates of the map ϕ in (3.3.4) are

$$
\begin{aligned}
p_{1111} &= a \cdot b_1 \cdot c_1 \cdot d_{11} \\
p_{1112} &= a \cdot b_1 \cdot c_1 \cdot (1 - d_{11}) \\
p_{1121} &= a \cdot b_1 \cdot (1 - c_1) \cdot d_{12} \\
p_{1122} &= a \cdot b_1 \cdot (1 - c_1) \cdot (1 - d_{12}) \\
p_{1211} &= a \cdot (1 - b_1) \cdot c_1 \cdot d_{21} \\
p_{1212} &= a \cdot (1 - b_1) \cdot c_1 \cdot (1 - d_{21}) \\
p_{1221} &= a \cdot (1 - b_1) \cdot (1 - c_1) \cdot d_{22} \\
p_{1222} &= a \cdot (1 - b_1) \cdot (1 - c_1) \cdot (1 - d_{22}) \\
p_{2111} &= (1 - a) \cdot b_2 \cdot c_2 \cdot d_{11} \\
p_{2112} &= (1 - a) \cdot b_2 \cdot c_2 \cdot (1 - d_{11}) \\
p_{2121} &= (1 - a) \cdot b_2 \cdot (1 - c_2) \cdot d_{12} \\
p_{2122} &= (1 - a) \cdot b_2 \cdot (1 - c_2) \cdot (1 - d_{12}) \\
p_{2211} &= (1 - a) \cdot (1 - b_2) \cdot c_2 \cdot d_{21} \\
p_{2212} &= (1 - a) \cdot (1 - b_2) \cdot c_2 \cdot (1 - d_{21}) \\
p_{2221} &= (1 - a) \cdot (1 - b_2) \cdot (1 - c_2) \cdot d_{22} \\
p_{2222} &= (1 - a) \cdot (1 - b_2) \cdot (1 - c_2) \cdot (1 - d_{22}).
\end{aligned}
$$

Note that each of the six quadrics above vanish under this specialization. In fact, the prime ideal of all algebraic relations among these sixteen quantities is

$$
I_{\text{local}(G)} + \left\langle \sum_{i=1}^{2} \sum_{j=1}^{2} \sum_{k=1}^{2} \sum_{l=1}^{2} p_{ijkl} - 1 \right\rangle.
$$

A detailed computer algebra study of discrete directed graphical models with at most five nodes was undertaken by Garcia, Stillman and Sturmfels in [50].   □

We now examine the case of multivariate normal random vectors. The recursive factorization (3.3.3) translates into a sequence of recursive regressions of random variables lower in the graph in terms of random variables farther up the graph. Indeed, suppose that the vertices of the graph G are ordered so that $j \to k$ is an edge only if $j < k$. For each $j \in [m]$, let $\varepsilon_j \sim \mathcal{N}(\nu_j, \omega_j^2)$ be a normal random variable and assume that $\varepsilon_1, \dots, \varepsilon_m$ are independent. To each edge $j \to k$ in the graph, we associate a *regression coefficient* λ_{jk}. We can then define a random vector $X = (X_1, \dots, X_m)$ as the solution to the recursive linear equations

$$X_k = \sum_{j \in \mathrm{pa}(k)} \lambda_{jk} X_j + \varepsilon_k, \quad k \in [m]. \tag{3.3.6}$$

The random vector X is multivariate normal, with a mean vector and covariance matrix whose entries are polynomial functions of the parameters ν_j, ω_j^2, and λ_{jk}. More precisely, $X \sim \mathcal{N}(\Lambda^{-T}\nu, \Lambda^{-T}\Omega\Lambda^{-1})$, where $\nu = (\nu_1, \dots, \nu_m)$, the matrix $\Omega = \mathrm{diag}(\omega_1^2, \dots, \omega_m^2)$ is diagonal, and the matrix Λ is upper-triangular with

$$\Lambda_{jk} = \begin{cases} 1 & \text{if } j = k, \\ -\lambda_{jk} & \text{if } j \to k \in E, \\ 0 & \text{otherwise.} \end{cases} \tag{3.3.7}$$

It can be shown, by induction on the number m of nodes, that the conditional distribution of X_k given $X_j = x_j$ for all $j < k$ is the normal distribution

$$\mathcal{N}\left(\nu_k + \sum_{j \in \mathrm{pa}(k)} \lambda_{jk} x_j, \omega_k^2\right).$$

This is also the conditional distribution of X_k given $X_j = x_j$ for all $j \in \mathrm{pa}(k)$. From these observations one can infer that the density $f(x)$ of a multivariate normal distribution $\mathcal{N}(\mu, \Sigma)$ factorizes as in (3.3.3) if and only if $\Sigma = \Lambda^{-T}\Omega\Lambda^{-1}$. In other words, the following result holds.

Proposition 3.3.12. *The parametrized Gaussian directed graphical model associated to G corresponds to all pairs $(\mu, \Sigma) \in \mathbb{R}^m \times PD_m$ such that $\Sigma = \Lambda^{-T}\Omega\Lambda^{-1}$ with Λ upper-triangular as in (3.3.7) and Ω diagonal with positive diagonal entries.*

Let $J_{\mathrm{global}(G)}$ be the ideal generated by all the determinantal constraints coming from the global Markov property of the DAG G, that is, those that come from the d-separation characterization in Definition 3.2.4. The ideal $J_{\mathrm{global}(G)}$ is generated by certain minors of the covariance matrix, spelled out explicitly in Definition 3.1.14. Let J_G be the vanishing ideal of all covariance matrices coming from the parametrization in Proposition 3.3.12. By definition, J_G is a prime ideal. The recursive factorization theorem (Theorem 3.3.2) guarantees that:

Proposition 3.3.13. *The set of positive definite matrices satisfying the conditional independence constraints equals the set of positive definite matrices that factorize as $\Sigma = \Lambda^{-T}\Omega\Lambda^{-1}$. In particular, the following equality of semi-algebraic sets holds:*

$$V(J_{\mathrm{global}(G)}) \cap PD_m = V(J_G) \cap PD_m.$$

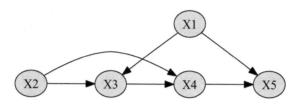

Figure 3.3.2: Directed acyclic graph, often called the Verma graph.

Unlike the discrete case, there can be probability distributions that satisfy all conditional independence constraints, but are not limits of Gaussian densities that factorize according to the graph. These will correspond to singular positive *semi-definite* covariance matrices, which yield probability distributions concentrated on lower-dimensional planes in $\mathbb{R}^m$. Here is an example where this happens:

Example 3.3.14 (Verma graph). Let G be the DAG on five vertices depicted in Figure 3.3.2. This graph is often called the Verma graph. The matrix Λ^{-1} in this case is the path polynomial matrix

$$
\Lambda^{-1} = \begin{pmatrix}
1 & 0 & \lambda_{13} & \lambda_{13}\lambda_{34} & \lambda_{13}\lambda_{34}\lambda_{45} + \lambda_{15} \\
0 & 1 & \lambda_{23} & \lambda_{23}\lambda_{34} + \lambda_{24} & \lambda_{23}\lambda_{34}\lambda_{45} + \lambda_{24}\lambda_{45} \\
0 & 0 & 1 & \lambda_{34} & \lambda_{34}\lambda_{45} \\
0 & 0 & 0 & 1 & \lambda_{45} \\
0 & 0 & 0 & 0 & 1
\end{pmatrix}.
$$

The conditional independence statements expressed by the *d*-separation relations in the Verma graph G are all implied by the three CI statements

$$
1 \perp\!\!\!\perp 2, \quad 1 \perp\!\!\!\perp 4 \mid \{2,3\}, \quad \text{and} \quad \{2,3\} \perp\!\!\!\perp 5 \mid \{1,4\}.
$$

Thus, the conditional independence ideal $J_{\mathrm{global}(G)}$ is generated by one linear form and five determinantal cubics. Using algebraic implicitization (e.g. in `Singular`), we find that $J_G = J_{\mathrm{global}(G)} + \langle f \rangle$ where f is the degree 4 polynomial

$$
\begin{aligned}
f &= \sigma_{23}\sigma_{24}\sigma_{25}\sigma_{34} - \sigma_{22}\sigma_{25}\sigma_{34}^2 - \sigma_{23}\sigma_{24}^2\sigma_{35} + \sigma_{22}\sigma_{24}\sigma_{34}\sigma_{35} \\
&\quad - \sigma_{23}^2\sigma_{25}\sigma_{44} + \sigma_{22}\sigma_{25}\sigma_{33}\sigma_{44} + \sigma_{23}^2\sigma_{24}\sigma_{45} - \sigma_{22}\sigma_{24}\sigma_{33}\sigma_{45}.
\end{aligned}
$$

The primary decomposition of $J_{\mathrm{global}(G)}$ is $J_G \cap \langle \sigma_{11}, \sigma_{12}, \sigma_{13}, \sigma_{14} \rangle$. Therefore, the zero set of $J_{\mathrm{global}(G)}$ inside the positive semi-definite cone contains singular covariance matrices that are not limits of covariance matrices that belong to the model. Note that since none of the indices of the σ_{ij} appearing in f contain 1, f vanishes on the marginal distribution for the random vector (X_2, X_3, X_4, X_5). This is the Gaussian version of what is known as the Verma constraint in machine learning; compare for example [77, §7.3.1]. The computation shows that the Verma constraint is still needed as a generator of the unmarginalized Verma model. $\square$

Chapter 4

Hidden Variables

One of the goals of algebraic statistics is to extend classical statistical theory for smooth statistical models to models that have singularities. Typically, these singularities arise in statistical models with hidden variables, in which a smooth model involving both observed and hidden variables is projected, via marginalization, to a model for the observed variables only. Hidden variable models are ubiquitous in statistical applications, but standard asymptotic theory usually does not apply because of model singularities. For example, we saw in Chapter 2 that the chi-square asymptotics for the likelihood ratio test are typically not valid at a singular point.

This chapter describes hidden variable models in some specific instances, and begins to address their geometric structure. Section 4.1 explains ties between hidden variable models and the algebraic geometry notion of secant varieties. The statistical models presented there are for discrete random variables. A Gaussian example, namely, the factor analysis model, is discussed in detail in Section 4.2.

4.1 Secant Varieties in Statistics

In this section, we focus on a special family of algebraic varieties called secant varieties. These varieties (or rather their positive parts) arise as special cases of the statistical models with hidden variables we are interested in studying. The hope is that by performing a detailed study of these particular models, much can be learned about more general families of mixtures and hidden variable models.

Let $\mathbb{K}$ be a field and consider two affine varieties $V, W \subseteq \mathbb{K}^k$. The *join* of V and W is the affine algebraic variety

$$\mathcal{J}(V, W) = \overline{\{\lambda v + (1 - \lambda)w \ : \ v \in V, w \in W, \lambda \in \mathbb{K}\}}.$$

This is the (Zariski) closure of the set of all points lying on lines spanned by a point in V and a point in W. If $V = W$ then this is the *secant variety* of V, denoted $\mathrm{Sec}^2(V) = \mathcal{J}(V, V)$. The *s-th higher secant variety* is defined by the rule:

$$\mathrm{Sec}^1(V) = V, \quad \mathrm{Sec}^s(V) = \mathcal{J}(\mathrm{Sec}^{s-1}(V), V).$$

In other words, the s-th secant variety is the closure of the set of all points on planes spanned by s points lying on V. For large s, this operation stabilizes, and we get the *affine hull* of V. If V and W are cones (that is, $\lambda v \in V$ for all $v \in V, \lambda \in \mathbb{K}$) then we can drop the affine combinations in the definition of join and merely take sums of elements. If V and W are projective varieties in $\mathbb{P}^{k-1}$ then their join $\mathcal{J}(V, W)$ is the projective variety associated with the join of their affine cones.

Secant varieties are well-studied objects in algebraic geometry. See [48, 104] for two monographs highlighting the connections of joins and secant varieties to classical algebraic geometry. Most of the algebraic geometry literature on secant varieties focuses on their dimensions, namely, to what extent this dimension can differ from the expected dimension. However, there is now also an emerging body of work on the vanishing ideals of secant varieties.

The statistical models we consider are generally not algebraic varieties, but are usually semi-algebraic sets. Furthermore, when we form new statistical models we typically take convex combinations of elements rather than affine combinations. This leads us to the notion of a mixture. For the rest of this section, we assume that our statistical models correspond, via their parameter spaces, to subsets of a real vector space $\mathbb{R}^k$. Let V and W be two such sets. Their *mixture* is the new set

$$\mathrm{Mixt}(V, W) = \{\lambda v + (1 - \lambda)w \mid v \in V, w \in W, \lambda \in [0, 1]\}.$$

The mixture $\mathrm{Mixt}(V, W)$ consists of all convex combinations of a point in V and a point in W. We define the higher mixtures, of a set with itself, by the rules

$$\mathrm{Mixt}^1(V) = V, \quad \mathrm{Mixt}^s(V) = \mathrm{Mixt}(\mathrm{Mixt}^{s-1}(V), V).$$

We can write the higher mixtures explicitly as follows:

$$\mathrm{Mixt}^s(V) = \left\{ \sum_{i=1}^{s} \lambda_i v_i \mid v_i \in V, \lambda_i \in [0, 1] \text{ and } \lambda_1 + \cdots + \lambda_s = 1 \right\}.$$

Note that for large s this operation stabilizes, and we obtain the convex hull of V.

Proposition 4.1.1. *If V is a semi-algebraic set, then the secant variety $\mathrm{Sec}^s(\overline{V})$ is the Zariski closure of the mixture $\mathrm{Mixt}^s(V)$.*

Proof. The mixture is clearly contained in the secant variety. If the λ_i are chosen generically in $[0, 1]$ then the corresponding point in $\mathrm{Mixt}^s(V)$ is a non-singular point in $\mathrm{Sec}^s(\overline{V})$. From this the assertion can be derived. $\square$

Secant varieties and mixtures can be quite different from each other:

Example 4.1.2 (Non-negative rank). Let V be the set of all non-negative $r \times c$ matrices with rank ≤ 1. This is the non-negative part of the cone over the Segre variety $\mathbb{P}^{r-1} \times \mathbb{P}^{c-1}$. The secant variety $\mathrm{Sec}^s(\overline{V})$ consists of all $r \times c$ matrices of rank $\leq s$. On the other hand, the mixture $\mathrm{Mixt}^s(V)$ consists of all non-negative $r \times c$ matrices that can be written as a sum of $\leq s$ non-negative matrices of rank 1.

The smallest number s such that a non-negative matrix A can be written as the sum of s non-negative matrices of rank 1 is called the *non-negative rank* of A. Thus $\mathrm{Mixt}^s(V)$ consists of all matrices of non-negative rank $\leq s$. To illustrate the differences between these two sets, we note that, whereas the rank of a matrix can be computed in polynomial time, determining its non-negative rank is NP-hard [95]. In other words, deciding whether a matrix $A \in \mathbb{Q}^{r \times c}$ belongs to $\mathrm{Sec}^s(\overline{V})$ can be decided in polynomial time, whereas deciding whether A belongs to $\mathrm{Mixt}^s(V)$ is unlikely to have a polynomial time algorithm. Specifically, it is known that

$$\mathrm{Mixt}^2(V) = \mathrm{Sec}^2(\overline{V}) \cap \mathbb{R}^{r \times c}_{\geq 0}$$
$$\text{but } \mathrm{Mixt}^s(V) \neq \mathrm{Sec}^s(\overline{V}) \cap \mathbb{R}^{r \times c}_{\geq 0} \text{ for } s \geq 3.$$

For instance, the following matrix has rank 3 but non-negative rank 4:

$$A = \begin{pmatrix} 1 & 1 & 0 & 0 \\ 0 & 1 & 1 & 0 \\ 0 & 0 & 1 & 1 \\ 1 & 0 & 0 & 1 \end{pmatrix}.$$

The set of all s-tuples of rank 1 matrices whose sum is a fixed matrix in $\mathrm{Mixt}^s(V)$ is known as the *space of explanations*. We refer to [71] for an interesting study of the topology of this space as well as a discussion of its statistical meaning. □

Now we are ready to describe two instances where mixtures arise in statistics.

Mixture models. The first family of models where a mixture arises are the discrete *mixture models*. To describe this family, suppose that $\mathcal{P} \subset \Delta_{r-1}$ is a statistical model for a random variable X with state space $[r]$. In the mixture model, we assume that there is a hidden random variable Y with state space $[s]$, and that for each $j \in [s]$, the conditional distribution of X given $Y = j$ is $p^{(j)} \in \mathcal{P}$. Furthermore, the random variable Y has some probability distribution $\pi \in \Delta_{s-1}$. Thus, the joint distribution of Y and X is given by the formula

$$P(Y = j, X = i) = \pi_j \cdot p_i^{(j)}.$$

However, we are assuming that Y is a hidden variable, so that we can only observe the marginal distribution of X, which is

$$P(X = i) = \sum_{j=1}^{s} \pi_j \cdot p_i^{(j)}.$$

In other words, the marginal distribution of X is the convex combination of the s distributions $p^{(1)}, \ldots, p^{(s)}$, with weights given by π. The mixture model consists of all probability distributions that can arise in this way.

Definition 4.1.3. Let $\mathcal{P} \subset \Delta_{r-1}$ be a statistical model. The *s-th mixture model* is

$$\mathrm{Mixt}^s(\mathcal{P}) \;=\; \left\{ \sum_{j=1}^s \pi_j p^{(j)} \;:\; \pi \in \Delta_{s-1} \text{ and } p^{(j)} \in \mathcal{P} \text{ for all } j \right\}.$$

Mixture models provide a way to build complex models out of simpler models. The basic assumption is that the underlying population to be modeled is one that can be split into s disjoint subpopulations. In each subpopulation, the underlying observable random variable X follows a probability distribution from the (simple) model $\mathcal{P}$. However, upon marginalization, the structure of the probability distribution is significantly more complex, because it is the convex combination of these simple probability distributions.

We have already seen a number of mixture models in the preceding chapters.

Example 4.1.4 (Non-negative rank, revisited). The independence model $\mathcal{M}_{X \perp\!\!\!\perp Y}$ is the set of rank 1 probability matrices, so that $\mathcal{M}_{X \perp\!\!\!\perp Y} = V \cap \Delta_{rc-1}$, where V is the set of non-negative matrices of rank ≤ 1 from Example 4.1.2. Thus the s-th mixture model $\mathrm{Mixt}^s(\mathcal{M}_{X \perp\!\!\!\perp Y})$ is the set of probability matrices of non-negative rank $\leq s$. As we argued in Example 4.1.2, this is a very complicated set if $s > 2$. □

Example 4.1.5 (The cheating coin flipper). To illustrate the discrepancy between the complex algebraic geometry and semi-algebraic geometry inherent in these mixture models, consider the model of the cheating coin flipper from Example 2.2.3. This is a mixture model with two hidden states of a binomial random variable with four trials. To simplify our analysis, suppose that the number of hidden states s is ≥ 4, so that our model $\mathrm{Mixt}^s(V)$ is the convex hull of the monomial curve

$$V = \left\{ ((1-\alpha)^4, 4\alpha(1-\alpha)^3, 6\alpha^2(1-\alpha)^2, 4\alpha^3(1-\alpha), \alpha^4) \;:\; \alpha \in [0,1] \right\}.$$

Among the semi-algebraic constraints of this convex hull are the conditions that the following two Hankel matrices are positive semi-definite:

$$\begin{pmatrix} 12p_0 & 3p_1 & 2p_2 \\ 3p_1 & 2p_2 & 3p_3 \\ 2p_2 & 3p_3 & 12p_4 \end{pmatrix} \succeq 0 \quad \text{and} \quad \begin{pmatrix} 3p_1 & 2p_2 \\ 2p_2 & 3p_3 \end{pmatrix} \succeq 0. \tag{4.1.1}$$

We drew $1,000,000$ random points according to a uniform distribution on the probability simplex Δ_4 and found that only $91,073$ satisfied these semi-algebraic constraints. Roughly speaking, the mixture model takes up only $\leq 10\%$ of the probability simplex, whereas the secant variety $\mathrm{Sec}^s(V)$ fills the simplex. We do not know whether the linear matrix inequalities in (4.1.1) suffice to characterize the mixture model, so it is possible that the model takes up an even smaller percentage of the probability simplex. □

Among the most important discrete mixture models is the *latent class model*, in which the underlying model $\mathcal{P}$ is the model of complete independence for m random variables $X_1, X_2, \ldots, X_m$. Here, complete independence $X_1 \perp\!\!\!\perp X_2 \perp\!\!\!\perp \ldots \perp\!\!\!\perp X_m$ means that the statement $X_A \perp\!\!\!\perp X_B$ holds for every partition $A \cup B = [m]$. A joint distribution belongs to the model of complete independence if and only if

$$p_i = \prod_{k=1}^{m} (p|_{\{k\}})_{i_k} \quad \text{for all } i \in \mathcal{R} = \prod_{j=1}^{m} [r_j].$$

In other words, this model consists of all probability distributions that are rank 1 tensors. Describing the mixture model of the complete independence model amounts to supposing that $X_1, X_2, \ldots, X_m$ are all conditionally independent, given the hidden variable Y. Passing to the Zariski closure, we are left with the problem of studying the secant varieties of the Segre varieties $\mathbb{P}^{r_1-1} \times \cdots \times \mathbb{P}^{r_m-1}$.

Proposition 4.1.6. *The mixture model* $\text{Mixt}^s(\mathcal{M}_{X_1 \perp\!\!\!\perp X_2 \perp\!\!\!\perp \cdots \perp\!\!\!\perp X_m})$ *consists of all probability distributions of* non-negative tensor rank *less than or equal to* s.

Given this interpretation as a conditional independence model with hidden variables, the mixture model of complete independence is also a graphical model with hidden variables (based on either a directed or an undirected graph). In the directed case, the graph has the edges $Y \to X_j$ for all j.

There are many important algebraic problems about latent class models, the solutions of which would be useful for statistical inference. By far the most basic, but still unanswered, problem is to determine the dimensions of these models. There has been much work on this problem, and in some situations, the dimensions of all secant varieties are known. For instance, if we only have two random variables X_1 and X_2, then the secant varieties are the classical determinantal varieties and their dimensions, and thus the dimensions of the mixture models, are all known. However, already in the case $m = 3$, it is an open problem to determine the dimensions of all the secant varieties, as s, r_1, r_2, and r_3 vary.

Example 4.1.7 (Identifiability of mixture models). Consider the mixture model $\text{Mixt}^2(\mathcal{M}_{X_1 \perp\!\!\!\perp X_2 \perp\!\!\!\perp X_3})$ where X_1, X_2, and X_3 are binary. A simple parameter count gives the *expected dimension* of the mixture model as $2 \times 3 + 1 = 7 = \dim \Delta_{\mathcal{R}-1}$. It is known, and the code below verifies, that this expected dimension is correct, and the mixture model is a full dimensional subset of probability simplex.

The next natural question to ask is: Is the model *identifiable*? Equivalently, given a probability distribution that belongs to the model, is it possible to recover the parameters in the probability specification. The following Macaulay2 code shows that the mixing parameter (labeled q) can be recovered by solving a quadratic equation in q whose coefficients are polynomials in the p_{ijk}. Note that the coefficient of q^2 is precisely the hyperdeterminant, featured in Example 2.2.10.

```
S = QQ[l, a,b,c,d,e,f,t];
R = QQ[q,p111,p112,p121,p122,p211,p212,p221,p222];
```

```
F = map(S,R,matrix{{
t*1,
t*(l*a*b*c + (1-l)*d*e*f),
t*(l*a*b*(1-c) + (1-l)*d*e*(1-f)),
t*(l*a*(1-b)*c + (1-l)*d*(1-e)*f),
t*(l*a*(1-b)*(1-c) + (1-l)*d*(1-e)*(1-f)),
t*(l*(1-a)*b*c + (1-l)*(1-d)*e*f),
t*(l*(1-a)*b*(1-c) + (1-l)*(1-d)*e*(1-f)),
t*(l*(1-a)*(1-b)*c + (1-l)*(1-d)*(1-e)*f),
t*(l*(1-a)*(1-b)*(1-c) + (1-l)*(1-d)*(1-e)*(1-f))}});
I = kernel F
```

The degree 2 that arises here shows the trivial non-identifiability called "population swapping" or "label switching" which amounts to the fact that in the mixture model we cannot tell the two subpopulations apart. The two solutions to this quadric will always be a pair $\lambda, 1 - \lambda$. The quadratic equation in q can also be used to derive some nontrivial semi-algebraic constraints for this mixture model. If there is a solution, it must be real, so the discriminant of this equation must be positive. This condition describes the real secant variety. Other semi-algebraic conditions arise by requiring that the two solutions lie in the interval $[0, 1]$.   □

Example 4.1.8 (A secant variety with dimension defect). Consider the mixture model $\text{Mixt}^3(\mathcal{M}_{X_1 \perp\!\!\!\perp X_2 \perp\!\!\!\perp X_3 \perp\!\!\!\perp X_4})$ where X_1, X_2, X_3 and X_4 are binary. The expected dimension of this model is $3 \times 4 + 2 = 14$, but it turns out that the true dimension is only 13. Indeed, the secant variety $\text{Sec}^3(\mathbb{P}^1 \times \mathbb{P}^1 \times \mathbb{P}^1 \times \mathbb{P}^1)$ is well-known to be defective. It is described implicitly by the vanishing of the determinants of the two 4×4 matrices:

$$
\begin{pmatrix}
p_{1111} & p_{1112} & p_{1121} & p_{1122} \\
p_{1211} & p_{1212} & p_{1221} & p_{1222} \\
p_{2111} & p_{2112} & p_{2121} & p_{2122} \\
p_{2211} & p_{2212} & p_{2221} & p_{2222}
\end{pmatrix}
\qquad
\begin{pmatrix}
p_{1111} & p_{1112} & p_{1211} & p_{1212} \\
p_{1121} & p_{1122} & p_{1221} & p_{1222} \\
p_{2111} & p_{2112} & p_{2211} & p_{2212} \\
p_{2121} & p_{2122} & p_{2221} & p_{2222}
\end{pmatrix}
$$

and thus is a complete intersection of degree 16 in $\mathbb{P}^{15}$.   □

Another problem, which is likely to require an even deeper investigation, is to understand the singularities of these mixture models. The importance of the singularities for statistical inference is a consequence of the following proposition.

Proposition 4.1.9. *Suppose that* $\text{Sec}^s(V) \neq \text{aff}(V)$, *the affine hull of* V. *Then*

$$\text{Sec}^{s-1}(V) \subseteq \text{Sing}(\text{Sec}^s(V)).$$

Proof. If f is any non-zero polynomial in the vanishing ideal $\mathcal{I}(\text{Sec}^s(V)) \subseteq \mathbb{C}[p]$, then any first-order partial derivative $\frac{\partial f}{\partial p_i}$ belongs to the ideal $\mathcal{I}(\text{Sec}^{s-1}(V))$. One way to prove this result is based on prolongations [84]. This approach implies that the Jacobian matrix associated to any generating set of $\mathcal{I}(\text{Sec}^s(V))$ evaluated at a point of $\text{Sec}^{s-1}(V)$ is the zero matrix.   □

Thus the study of the behavior of the likelihood ratio test statistic, as in Section 2.3, for hypothesis testing with respect to the models $\mathrm{Mixt}^{s-1}(\mathcal{P}) \subset \mathrm{Mixt}^s(\mathcal{P})$ will require a careful analysis of the singularities of these secant varieties.

Phylogenetic models. Another important family of statistical models involving mixtures, and thus secant varieties, arises in phylogenetics (see [46, 83] for book-length introductions to this area). Phylogenetic models are graphical models with hidden variables that are defined over trees. The nodes in the tree correspond to a site in aligned DNA sequences of several species. The leaves of the trees represent species that are alive today and whose DNA is available for analysis. Internal nodes in the tree correspond to extinct ancestral species whose DNA is not available. Thus, all internal nodes of the tree correspond to hidden random variables.

Typically, we assume that each discrete random variable represented in the tree has the same number of states r. When working with the nucleotides directly, this number is 4. If we compress our DNA consideration to only look at mutations across the purine/pyrimidine divide, then each random variable would have only two states. Going to the other extreme, if we consider regions of DNA that code for proteins, we could group the DNA into codons that correspond to one of twenty amino acids. Here we will focus primarily on the case of either $r = 2$ or $r = 4$.

A particular phylogenetic model is specified by placing restrictions on the *transition matrices* that can be used on the edges in the tree. A transition matrix contains the conditional probabilities for a random variable given its (unique) parent variable in the tree. The largest possible model, allowing the biggest possible class of transition structures, is known as the *general Markov model*. In the general Markov model, the transition matrices are unconstrained except that they should actually contain valid conditional probabilities. However, one often also considers other classes of models that are submodels of the general Markov model.

One of the basic problems of *phylogenetic algebraic geometry* is to determine the vanishing ideals of phylogenetic models. Given a tree T, and particular choice of transition structure, we get a rational map ϕ_T from a low dimensional parameter space (the space of all suitably structured transition matrices) into the high dimensional probability simplex containing the probability distributions for the random variables at the m leaves. The image of this map $\mathrm{im}(\phi_T)$ is a semi-algebraic set in Δ_{r^m-1}, and we would like to determine its vanishing ideal $I_T = \mathcal{I}(\mathrm{im}(\phi_T)) \subseteq \mathbb{R}[p]$.

A fundamental result of Draisma and Kuttler says that for "reasonable" algebraic phylogenetic models, the problem of determining a generating set of the phylogenetic ideals I_T for an arbitrary tree T can be reduced, via a combinatorial procedure, to very small trees. We here do not offer a formal definition of what "reasonable" means but refer to [36] instead. Let $K_{1,m}$ denote the complete bipartite graph with one hidden node and m observed nodes. These graphs are often called *claws*.

Theorem 4.1.10 (Draisma-Kuttler [36]). *Given a "reasonable" phylogenetic model, there is an explicit combinatorial procedure to build generators for the phylogenetic* *ideal I_T from the generators of the phylogenetic ideal $I_{K_{1,m}}$, where m is the largest degree of a vertex in T.*

For the rest of this section, we focus exclusively on the case of phylogenetic models on claw trees. For many classes of transition matrices used in phylogenetics, the algebraic varieties corresponding to claws are secant varieties of toric varieties.

There are three important situations to which the Draisma-Kuttler theorem applies. First, and most classically, is the class of group-based phylogenetic models. After applying the Fourier transform, these models decompose into simple cases of claw trees. We assume that the underlying trees are *bifurcating* (or trivalent or binary, depending on the reference), in which case we only need to understand the model for three-leaf claw trees. The vanishing ideal for three-leaf claw trees can be determined by a computation, first carried out in [89]. We refer the reader to this paper for the details on group based models, and the resulting toric structure.

The next important case where the Draisma-Kuttler theorem is applicable is the *general Markov model* mentioned above. See also [3] and [73, §19.1]. To solve the case of bifurcating trees, we must again determine the vanishing ideal for the three-leaf claw tree. If there are r states for the random variables, we must determine the vanishing ideal of the parametrization

$$\phi : \mathbb{R}^r \times \mathbb{R}^{r^2} \times \mathbb{R}^{r^2} \times \mathbb{R}^{r^2} \longrightarrow \mathbb{R}^{r^3}$$

$$(\pi, A, B, C) \mapsto \sum_{i=1}^{r} \pi_i \cdot A_{i \cdot} \otimes B_{i \cdot} \otimes C_{i \cdot} .$$

In phylogenetic models, the root distribution parameter π is in the simplex Δ_{r-1}. For each fixed value $i \in [r]$, the tensor $A_{i \cdot} \otimes B_{i \cdot} \otimes C_{i \cdot} = (a_{ij}b_{ik}c_{il})$ has rank 1, and hence belongs to the model of complete independence $\mathcal{M}_{X_1 \perp\!\!\!\perp X_2 \perp\!\!\!\perp X_3}$.

Proposition 4.1.11. *The general Markov model on a three leaf claw tree for r states is the same family of probability distributions as the mixture model*

$$\mathrm{Mixt}^r(\mathcal{M}_{X_1 \perp\!\!\!\perp X_2 \perp\!\!\!\perp X_3})$$

where each X_i has r states. The projectivized Zariski closure of the model is the secant variety

$$\mathrm{Sec}^r(\mathbb{P}^{r-1} \times \mathbb{P}^{r-1} \times \mathbb{P}^{r-1}).$$

The secant variety $\mathrm{Sec}^2(\mathbb{P}^1 \times \mathbb{P}^1 \times \mathbb{P}^1)$ fills all of projective space $\mathbb{P}^7$; recall Example 4.1.7. Therefore, the vanishing ideal of the general Markov model on a bifurcating tree with binary states can be described very explicitly:

Theorem 4.1.12. *Let T be a bifurcating tree, and let each random variable be binary $(r = 2)$. Then the phylogenetic ideal I_T for the general Markov model is generated by the 3×3 minors of all flattenings of the table $(p_{i_1 \ldots i_m})$ that come from splits in the tree T.*

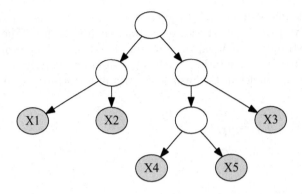

Figure 4.1.1: Bifurcating tree.

Proof. This was conjectured in [50, §7] and proved in [5] and in [65]. The Draisma-Kuttler theorem [36] can be regarded as a generalization of this statement. □

Example 4.1.13. Let T be the bifurcating tree with five leaves in Figure 4.1.1. Let

$$
P_{12|345} = \begin{pmatrix}
p_{11111} & p_{11112} & p_{11121} & p_{11122} & p_{11211} & p_{11212} & p_{11221} & p_{11222} \\
p_{12111} & p_{12112} & p_{12121} & p_{12122} & p_{12211} & p_{12212} & p_{12221} & p_{12222} \\
p_{21111} & p_{21112} & p_{21121} & p_{21122} & p_{21211} & p_{21212} & p_{21221} & p_{21222} \\
p_{22111} & p_{22112} & p_{22121} & p_{22122} & p_{22211} & p_{22212} & p_{22221} & p_{22222}
\end{pmatrix}
$$

and

$$
P_{123|45} = \begin{pmatrix}
p_{11111} & p_{11112} & p_{11121} & p_{11122} \\
p_{11211} & p_{11212} & p_{11221} & p_{11222} \\
p_{12111} & p_{12112} & p_{12121} & p_{12122} \\
p_{12211} & p_{12212} & p_{12221} & p_{12222} \\
p_{21111} & p_{21112} & p_{21121} & p_{21122} \\
p_{21211} & p_{21212} & p_{21221} & p_{21222} \\
p_{22111} & p_{22112} & p_{22121} & p_{22122} \\
p_{22211} & p_{22212} & p_{22221} & p_{22222}
\end{pmatrix}.
$$

These two matrices correspond to the two non-trivial splits of the tree T. The phylogenetic ideal for $r = 2$ is generated by all 3×3 minors of the two matrices. □

For $r = 3$, there is also an explicit description in the case of bifurcating trees, which follows from Theorem 4.1.10 and [66]. However, for the most interesting case of DNA sequences ($r = 4$), it is still an open problem to describe the generating sets of these ideals. It is known that certain polynomials of degrees 5, 6, and 9 are needed as generators, but it is unknown whether these polynomials suffice [65, 88].

Another case where an interesting secant variety appears is the *strand symmetric model* (see Chapter 16 in [73]). In this model, we make restrictive assumptions on the allowable transition matrices. The restrictions are based on the fact that DNA is double-stranded, and the strands always form base pairs. In particular, A is always paired with T and C is always paired with G. The base pairing of the double stranded DNA sequence means that a mutation on one strand will always force a balancing mutation on the opposite strand. We infer the following equalities between entries in a transition matrix in the strand symmetric model:

$$\theta_{AA} = \theta_{TT}, \qquad\qquad \theta_{AC} = \theta_{TG},$$
$$\theta_{AG} = \theta_{TC}, \qquad\qquad \theta_{AT} = \theta_{TA},$$
$$\theta_{CA} = \theta_{GT}, \qquad\qquad \theta_{CC} = \theta_{GG},$$
$$\theta_{CG} = \theta_{GC}, \qquad\qquad \theta_{CT} = \theta_{GA}.$$

Rearranging the rows and columns of the constrained transition matrix θ, we see that it has the block form:

$$\theta = \begin{pmatrix} \alpha & \beta \\ \beta & \alpha \end{pmatrix} \quad \text{where} \quad \alpha = \begin{pmatrix} \theta_{AA} & \theta_{AG} \\ \theta_{GA} & \theta_{GG} \end{pmatrix}, \quad \beta = \begin{pmatrix} \theta_{AT} & \theta_{AC} \\ \theta_{GT} & \theta_{GC} \end{pmatrix}.$$

Furthermore, the strand symmetric assumption implies that the root distribution should satisfy the relationship $\pi_A = \pi_T$ and $\pi_C = \pi_G$. This block form implies that it is possible to use a *Fourier transform* to simplify the parametrization. In the Fourier coordinates, the underlying algebraic variety of the strand symmetric model has a simple combinatorial structure.

Definition 4.1.14. Let $\phi : \mathbb{P}^{r_1-1} \times \mathbb{P}^{r_2-1} \times \mathbb{P}^{r_3-1} \longrightarrow \mathbb{P}^{r_1 r_2 r_3 - 1}$ be the rational map

$$\phi_{ijk}(a, b, c) = \begin{cases} a_i b_j c_k & \text{if } i + j + k \text{ is even,} \\ 0 & \text{if } i + j + k \text{ is odd.} \end{cases}$$

The image of ϕ is the *checkerboard Segre variety*:

$$\mathrm{Seg}_{\mathbb{Z}_2}(\mathbb{P}^{r_1-1} \times \mathbb{P}^{r_2-1} \times \mathbb{P}^{r_3-1}).$$

The checkerboard Segre variety is a toric variety, and its vanishing ideal is generated by quadrics. Its secant varieties arise as the Zariski closure of the strand symmetric model.

Proposition 4.1.15 ([73, Chap. 16]). *The projectivized Zariski closure of the strand symmetric model for DNA sequences on the 3-leaf claw tree is the secant variety of the checkboard Segre variety:*

$$\mathrm{Sec}^2(\mathrm{Seg}_{\mathbb{Z}_2}(\mathbb{P}^3 \times \mathbb{P}^3 \times \mathbb{P}^3)).$$

While some of the equations in the ideal of this secant variety are known (in particular, equations of degree 3 and 4), it is still an open problem to compute its prime ideal. Once this problem has been solved, we could apply Theorem 4.1.10 to recover all equations for the strand symmetric model on any trivalent tree.

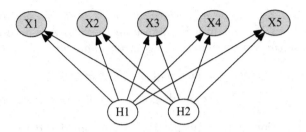

Figure 4.2.1: Directed acyclic graph for the factor analysis model $\mathcal{F}_{5,2}$.

4.2 Factor Analysis

Let $X_1, \ldots, X_m$ be a collection of continuous random variables that represent a randomly selected individual's performance when solving m math problems. It is natural to expect the individual to do well (or poorly) in most tasks provided he/she has (or does not have) a talent for the type of considered problems. Therefore, if the population the individual is selected from indeed exhibits varying mathematical talent, then the random variables $X_1, \ldots, X_m$ will be dependent. However, one might expect the variables $X_1, \ldots, X_m$ to be independent if the random selection of an individual occurs conditionally on a fixed talent level.

Situations of the type just described are the subject of *factor analysis*, where notions such as mathematical talent are quantified using one or more hidden variables $H_1, \ldots, H_s$. The hidden variables are termed *factors* and we assume throughout that their number is smaller than the number of observed variables, that is, $s < m$. The *factor analysis model* $\mathcal{F}_{m,s}$ for the joint distribution of $X_1, \ldots, X_m$ assumes that the random vector $(X_1, \ldots, X_m, H_1, \ldots, H_s)$ follows a joint multivariate normal distribution with a positive definite covariance matrix such that

$$X_1 \perp\!\!\!\perp X_2 \perp\!\!\!\perp \ldots \perp\!\!\!\perp X_m \,|\, (H_1, \ldots, H_s). \tag{4.2.1}$$

Display (4.2.1) refers to complete conditional independence of $X_1, \ldots, X_m$ given $H_1, \ldots, H_s$, that is,

$$X_A \perp\!\!\!\perp X_B \,|\, (H_1, \ldots, H_s) \quad \text{for all partitions } (A, B) \text{ of } [m]. \tag{4.2.2}$$

Having assumed a joint multivariate normal distribution, (4.2.2) is equivalent to

$$X_i \perp\!\!\!\perp X_j \,|\, (H_1, \ldots, H_s) \quad \text{for } 1 \leq i < j \leq s. \tag{4.2.3}$$

This equivalence is false for discrete random variables. We remark that the model $\mathcal{F}_{m,s}$ is a graphical model with hidden variables based on a directed acyclic graph that is complete bipartite with edges pointing from the hidden to the observed variables; recall Section 3.2. The graph for $\mathcal{F}_{5,2}$ is shown in Figure 4.2.1.

We start out by deriving the following parametric model representation from the determinantal implicit description given in (4.2.1) and (4.2.3).

Proposition 4.2.1. *The factor analysis model $\mathcal{F}_{m,s}$ is the family of multivariate normal distributions $\mathcal{N}_m(\mu, \Sigma)$ on $\mathbb{R}^m$ whose mean vector μ is an arbitrary vector in $\mathbb{R}^m$ and whose covariance matrix Σ lies in the (non-convex) cone*

$$F_{m,s} = \{\Omega + \Lambda\Lambda^T \in \mathbb{R}^{m\times m} : \Omega \succ 0 \text{ diagonal}, \Lambda \in \mathbb{R}^{m\times s}\}$$
$$= \{\Omega + \Psi \in \mathbb{R}^{m\times m} : \Omega \succ 0 \text{ diagonal}, \Psi \succeq 0 \text{ symmetric}, \text{rank}(\Psi) \leq s\}.$$

Here the notation $A \succ 0$ means that A is a positive definite matrix, and similarly $A \succeq 0$ means that A is a positive semi-definite matrix.

Proof. Consider the joint covariance matrix of hidden and observed variables,

$$\text{Cov}\begin{pmatrix} X \\ H \end{pmatrix} = \begin{pmatrix} \Sigma & \Lambda \\ \Lambda^T & \Phi \end{pmatrix}. \tag{4.2.4}$$

Using Definition 3.1.14, the conditional independence relations (4.2.3) translate into the vanishing of the following $(s+1) \times (s+1)$-determinants:

$$\det\begin{pmatrix} \sigma_{ij} & \Lambda_{i*} \\ \Lambda_{j*}^T & \Phi \end{pmatrix} = \det(\Phi) \cdot (\sigma_{ij} - \Lambda_{i*}\Phi^{-1}\Lambda_{j*}^T) = 0. \tag{4.2.5}$$

Here we assume $i \neq j$. Since $\det(\Phi) > 0$, (4.2.5) implies that the positive definite Schur complement $\Omega = \Sigma - \Lambda\Phi^{-1}\Lambda^T$ is diagonal. By Cholesky decomposition of Φ^{-1}, the covariance matrix $\Sigma = \Omega + \Lambda\Phi^{-1}\Lambda^T$ for the observed variables is seen to be in $F_{m,s}$, and all matrices in $F_{m,s}$ can be obtained in this fashion. $\qquad\square$

In what follows we identify the factor analysis model $\mathcal{F}_{m,s}$ with its parameter space $F_{m,s}$. By Proposition 4.2.1, the semi-algebraic set $F_{m,s}$ can be parametrized by the polynomial map with coordinates

$$\sigma_{ij} = \begin{cases} \omega_{ii} + \sum_{r=1}^{s} \lambda_{ir}^2 & \text{if } i = j, \\ \sum_{r=1}^{s} \lambda_{ir}\lambda_{jr} & \text{if } i < j, \end{cases} \tag{4.2.6}$$

where $\omega_{ii} > 0$ and $\lambda_{ij} \in \mathbb{R}$. Note that this parametrization can also be derived from Proposition 3.3.13.

The *dimension* $d = \dim(F_{m,s})$ of the model $F_{m,s}$ is equal to the maximal rank of the Jacobian matrix of the parametrization (4.2.6). The *codimension* of $F_{m,s}$ is $\binom{m+1}{2} - d$. The following result appears in [40, Thm. 2].

Theorem 4.2.2. *The dimension and the codimension of the factor analysis model are*

$$\dim(F_{m,s}) = \min\left\{ m(s+1) - \binom{s}{2}, \binom{m+1}{2} \right\},$$

$$\text{codim}(F_{m,s}) = \max\left\{ \binom{m-s}{2} - s, 0 \right\}.$$

From (4.2.6) it is evident that a matrix Σ is in $F_{m,s}$ if and only if it is the sum of s matrices in $F_{m,1}$. The sets $F_{m,s}$ being cones, the latter holds if and only if Σ is a convex combination of matrices in $F_{m,1}$. We observe the following structure.

Remark 4.2.3. *In factor analysis, $F_{m,s} = \mathrm{Mixt}^s(F_{m,1})$, and the Zariski closure of $F_{m,s}$ is the s-th secant variety of the Zariski closure of $F_{m,1}$.*

Such secant structure also arises in other Gaussian graphical models with hidden variables [91, §7.3]. For factor analysis, it implies that the singular locus of the factor analysis model with s factors contains the model with $s-1$ factors; recall Proposition 4.1.9. Therefore, the statistical estimation of the number of hidden factors presents a non-standard problem. It should be noted, however, that $F_{m,1}$ already has singularities in the cone of positive definite matrices, and that $F_{m,s}$ contains positive definite singularities outside $F_{m,s-1}$. The occurrence of singularities is related to the identifiability issues discussed in [6].

In the remainder of this section we will be interested in polynomial relations among the entries of a factor analysis covariance matrix $\Sigma \in F_{m,s}$. Let

$$I_{m,s} \;=\; \{f \in \mathbb{R}[\sigma_{ij},\, i \le j] \;:\; f(\Sigma) = 0 \text{ for all } \Sigma \in F_{m,s}\} \tag{4.2.7}$$

be the ideal of these relations. The ideal $I_{m,s}$ contains much useful information about the geometry of the model. But relations in $I_{m,s}$ can also serve as closed-form test statistics for what are commonly termed Wald tests. The next proposition follows from the asymptotic normality of the sample covariance matrix S and an application of the delta-method, which refers to using a Taylor-expansion in order to derive the asymptotic distribution of a transformation of S; compare [40, §3].

Proposition 4.2.4. *Let S be the sample covariance matrix of an n-sample drawn from a distribution $\mathcal{N}(\mu, \Sigma)$ in $\mathcal{F}_{m,s}$. Let $f \in I_{m,s}$ and $\mathrm{Var}_{\Sigma}[f(S)]$ the variance of sample evaluation. If the gradient df of the polynomial f is non-zero at Σ, then*

$$\frac{f(S)^2}{\mathrm{Var}_S[f(S)]} \;\xrightarrow{\; D \;}\; \chi_1^2 \qquad \text{as } n \to \infty.$$

The convergence in distribution in Proposition 4.2.4 justifies (asymptotic) p-value calculations. As stated the convergence result is most useful for hypersurfaces but it can be generalized to the case where several polynomials in $I_{m,s}$ are considered. It should be noted, however, that the validity of the χ^2-asymptotics is connected to the smoothness of the set $F_{m,s}$.

As we will see next, the $(s+1) \times (s+1)$-minors of Σ will play a particular role for the factor analysis ideal $I_{m,s}$. Their sampling variance and covariance structure is derived in [39]. Since membership in the Zariski closure of $F_{m,s}$ depends only on the off-diagonal entries of the matrix Σ, the ideal $I_{m,s}$ can be computed by elimination of the diagonal entries σ_{ii}:

Proposition 4.2.5 ([40, Thm. 7]). *Let $M_{m,s} \subseteq \mathbb{R}[\sigma_{ij},\, i \le j]$ be the ideal generated by all $(s+1) \times (s+1)$-minors of a symmetric matrix $\Sigma \in \mathbb{R}^{m \times m}$. Then*

$$I_{m,s} \;=\; M_{m,s} \cap \mathbb{R}[\sigma_{ij},\, i < j].$$

If the symmetric matrix $\Sigma \in \mathbb{R}^{m \times m}$ is of size $m \geq 2(s+1)$, then it contains $(s+1) \times (s+1)$-minors in $\mathbb{R}[\sigma_{ij}, \ i < j]$. Such *off-diagonal minors* are clearly in $I_{m,s}$. Each off-diagonal minor $\det(\Sigma_{A,B})$ is derived from two disjoint subsets $A, B \subset [m]$ of equal cardinality $s+1$, and thus, up to sign change, there are

$$\frac{1}{2}\binom{m}{2(s+1)}\binom{2(s+1)}{s+1}$$

off-diagonal minors.

Example 4.2.6 (Tetrad). Up to sign change, the matrix

$$\Sigma = \begin{pmatrix} \sigma_{11} & \sigma_{12} & \sigma_{13} & \sigma_{14} \\ \sigma_{12} & \sigma_{22} & \sigma_{23} & \sigma_{24} \\ \sigma_{13} & \sigma_{23} & \sigma_{33} & \sigma_{34} \\ \sigma_{14} & \sigma_{24} & \sigma_{34} & \sigma_{44} \end{pmatrix}$$

contains three off-diagonal 2×2-minors, namely

$$\sigma_{12}\sigma_{34} - \sigma_{13}\sigma_{24}, \quad \sigma_{14}\sigma_{23} - \sigma_{13}\sigma_{24}, \quad \sigma_{12}\sigma_{34} - \sigma_{14}\sigma_{23}. \tag{4.2.8}$$

In the statistical literature these minors are known as *tetrads* or *tetrad differences*, as they arise in the one-factor model with four observed variables [55]. The three tetrads in (4.2.8) are algebraically dependent: the third tetrad is the difference of the first and the second tetrad. In fact, the first and second tetrad generate the entire ideal $I_{4,1}$, as can be verified using the following computation in **Singular**:

```
LIB "elim.lib";
ring R = 0,(s11,s22,s33,s44, s12,s23,s34,s14, s13,s24),lp;
matrix S[4][4] = s11,s12,s13,s14,
                 s12,s22,s23,s24,
                 s13,s23,s33,s34,
                 s14,s24,s34,s44;
ideal M41 = minor(S,2);
eliminate(M41,s11*s22*s33*s44);
```

The command `eliminate` computes the intersection $M_{4,1} \cap \mathbb{R}[\sigma_{ij}, \ i < j]$ in Proposition 4.2.5 and thus (a finite generating set of) the ideal $I_{4,1}$. $\qquad \square$

The ideal structure encountered in Example 4.2.6 generalizes to larger models with $s = 1$ factor. Any four indices $i < j < k < \ell$ in $[m]$ define a 4×4-principal submatrix of Σ from which we can extract two algebraically independent tetrads. Choosing these two tetrads in the way the first two tetrads in (4.2.8) were obtained, we create the $2\binom{m}{4}$ tetrads

$$\mathcal{T}_m = \{\underline{\sigma_{ij}\sigma_{k\ell}} - \sigma_{ik}\sigma_{j\ell}, \ \underline{\sigma_{i\ell}\sigma_{jk}} - \sigma_{ik}\sigma_{j\ell} \mid 1 \leq i < j < k < \ell \leq m\}.$$

As described in [30], the underlined terms are the leading terms with respect to certain circular monomial orders on $\mathbb{R}[\sigma_{ij}, \ i < j]$. Let

$$d(i,j) = \min\{ (i - j \mod m), \ (j - i \mod m)\}$$

be the circular distance of two indices $i, j \in [m]$. Under a circular monomial order, $\sigma_{ij} \succ \sigma_{kl}$ if $d(i,j) < d(k,l)$. One example of a circular monomial order is the lexicographic order that was used in the **Singular** code in Example 4.2.6.

Theorem 4.2.7 ([30, Thm. 2.1]). *If $m \leq 3$ the ideal $I_{m,1}$ is the zero ideal. If $m \geq 4$, the set $\mathcal{T}_m$ is the reduced Gröbner basis of the ideal $I_{m,1}$ with respect to a circular monomial order.*

Example 4.2.8 (Pentad). If $m = 5$ and $s = 2$, then there are no off-diagonal 3×3-minors in $I_{5,2}$. Nevertheless, Theorem 4.2.2 informs us that $F_{5,2}$ has codimension 1, that is, it is a hypersurface in the space of symmetric 5×5-matrices. Adapting the **Singular** code presented in Example 4.2.6, we can compute this hypersurface. We find that the ideal $I_{5,2}$ is generated by the irreducible polynomial

$$\sigma_{12}\sigma_{13}\sigma_{24}\sigma_{35}\sigma_{45} - \sigma_{12}\sigma_{13}\sigma_{25}\sigma_{34}\sigma_{45} - \sigma_{12}\sigma_{14}\sigma_{23}\sigma_{35}\sigma_{45} + \sigma_{12}\sigma_{14}\sigma_{25}\sigma_{34}\sigma_{35}$$

$$+\sigma_{12}\sigma_{15}\sigma_{23}\sigma_{34}\sigma_{45} - \sigma_{12}\sigma_{15}\sigma_{24}\sigma_{34}\sigma_{35} + \sigma_{13}\sigma_{14}\sigma_{23}\sigma_{25}\sigma_{45} - \sigma_{13}\sigma_{14}\sigma_{24}\sigma_{25}\sigma_{35}$$

$$-\sigma_{13}\sigma_{15}\sigma_{23}\sigma_{24}\sigma_{45} + \sigma_{13}\sigma_{15}\sigma_{24}\sigma_{25}\sigma_{34} - \sigma_{14}\sigma_{15}\sigma_{23}\sigma_{25}\sigma_{34} + \sigma_{14}\sigma_{15}\sigma_{23}\sigma_{24}\sigma_{35}.$$

This polynomial is referred to as the *pentad* in the statistical literature. It was first derived in the 1930s by Kelley [64].

Why is the pentad in $I_{5,2}$? We can argue this by selecting two 3×3-minors that both involve exactly one element of the diagonal of Σ. For instance, consider the $\{1, 2, 3\} \times \{3, 4, 5\}$ and the $\{2, 3, 4\} \times \{1, 3, 5\}$-minors. We can expand these determinants as

$$\det (\Sigma_{123,345}) = \sigma_{33} \cdot a_{11}(\Sigma) + a_{10}(\Sigma), \tag{4.2.9a}$$

$$\det (\Sigma_{234,135}) = \sigma_{33} \cdot a_{21}(\Sigma) + a_{20}(\Sigma), \tag{4.2.9b}$$

where a_{k1} and a_{k0} are quadratic and cubic polynomials in $\mathbb{R}[\sigma_{ij}, \ i < j]$, respectively. Setting the minors in (4.2.9a) and (4.2.9b) equal to zero we obtain two equations, which we view as linear equations in the unknown σ_{33} with coefficients a_{kl}. Now recall that Proposition 4.2.1 states that a positive definite matrix Σ is in $F_{m,s}$ if and only if we can create a matrix of rank s by subtracting positive reals from the diagonal of Σ. Therefore, if the coefficients a_{kl} are derived from a matrix $\Sigma \in F_{5,2}$, then the two equations given by (4.2.9a) and (4.2.9b) have a solution in σ_{33}. This requires the following determinant to vanish:

$$\det \begin{pmatrix} a_{10}(\Sigma) & a_{11}(\Sigma) \\ a_{20}(\Sigma) & a_{21}(\Sigma) \end{pmatrix} = 0. \tag{4.2.10}$$

Plugging the original quadratic and cubic coefficient polynomials $a_{kl} \in \mathbb{R}[\sigma_{ij}, \ i < j]$ from (4.2.9a) and (4.2.9b) into (4.2.10) we obtain a quintic polynomial in $I_{5,2}$. Upon expansion of the determinant, this polynomial is seen to be the pentad. $\qquad\square$

The reasoning leading to (4.2.10) can be generalized. Suppose m and s are such that the codimension of $F_{m,s}$ is positive and $m \geq 2(s+1) - k$ for some $k \geq 1$. Then we can select a set of k indices $D = \{d_1 \ldots, d_k\} \subset [m]$ and $k + 1$ many $(s + 1) \times (s + 1)$-minors that all involve the diagonal entries σ_{ii}, $i \in D$, but no other diagonal entries. The expansion of the minors yields a system of $k + 1$ multilinear equations in the k unknowns σ_{ii}, $i \in D$. The other terms in the expansion are treated as coefficients for the equations. If these coefficients are derived from a matrix $\Sigma \in F_{m,s}$ then the multilinear equation system has a solution. It can be shown that this requires an irreducible polynomial in the coefficients of the system to vanish; compare [40, §6]. This polynomial is known as the k-th *multilinear resultant*. The determinant in (4.2.10) is the multilinear resultant for $k = 1$. Multilinear resultants can be employed to compute polynomials in $I_{m,s}$ in the same way as the determinant in (4.2.10) yields the pentad. In [40], this approach was used in particular to show that the Zariski closure of $F_{9,5}$ is a hypersurface defined by an irreducible homogeneous polynomial of degree 54.

Let us now return to the ideals $I_{m,s}$ for $s = 2$ factors. We have encountered two types of polynomials in $I_{m,2}$, namely, off-diagonal 3×3-minors and the pentad of degree 5. In Example 4.2.8 we have seen that the pentad generates $I_{5,2}$. The Gröbner basis computation underlying this result is also feasible for larger models. The following conjecture can be verified computationally for small to moderate m. It holds at least for $m \leq 9$.

Conjecture 4.2.9. *The ideal of the two-factor model, $I_{m,2}$, is minimally generated by* $5\binom{m}{6}$ *off-diagonal 3×3-minors and $\binom{m}{5}$ pentads.*

A Gröbner basis for $I_{m,2}$ is described in [92]. This Gröbner basis contains the conjectured minimal generating set but also additional polynomials of every odd degree between 3 and m. We refer to [40, Conjecture 28] for a specific conjecture about the case of $s = 3$ factors, based on various computational experiments.

We close this section by commenting on a particular symmetry structure in factor analysis models. Let $\binom{[m]}{k}$ be the set of all subsets $A \subset [m]$ that have cardinality k. If $A \in \binom{[m]}{k}$, then we write $I_{A,s}$ to denote the ideal $I_{k,s}$ when the entries of the submatrix $\Sigma_{A,A}$ are used as indeterminates. In Theorem 4.2.7, we have seen that if $m \geq 4$ then a generating set of the ideal $I_{m,1}$ can be obtained by taking the union of generating sets of the ideals $I_{A,1}$ for subsets $A \in \binom{[m]}{4}$. Similarly, if $m \geq 6$, then the generating set for $I_{m,2}$ proposed in Conjecture 4.2.9 is composed of generating sets of the ideals $I_{A,2}$ for $A \in \binom{[m]}{6}$. This raises the question whether such *finiteness* up to symmetry holds more generally; see Problem 7.8. A positive answer to this question would be important for statistical practice, as a statistical test of the model $\mathcal{F}_{m,s}$ for large m could be carried out by testing lower-dimensional models $\mathcal{F}_{A,s}$ for an appropriately chosen set of margins $A \subseteq [m]$.

Chapter 5

Bayesian Integrals

A key player in Bayesian statistics is the integrated likelihood function of a model for given data. The integral, also known as the marginal likelihood, is taken over the model's parameter space with respect to a probability measure that quantifies prior belief. While Chapter 2 was concerned with maximizing the likelihood function, we now seek to integrate that same function. This chapter aims to show how algebraic methods can be applied to various aspects of this problem. Section 5.1 discusses asymptotics of Bayesian integrals for large sample size, while Section 5.2 concerns exact evaluation of integrals for small sample size.

5.1 Information Criteria and Asymptotics

We fix a statistical model $\mathcal{P}_{\Theta} = \{P_{\theta} : \theta \in \Theta\}$ with parameter space $\Theta \subseteq \mathbb{R}^k$. Consider a sample of independent random vectors,

$$X^{(1)}, \ldots, X^{(n)} \sim P_{\theta_0}, \tag{5.1.1}$$

drawn from an (unknown) *true distribution* P_{θ_0} where $\theta_0 \in \Theta$. We say that a submodel given by a subset $\Theta_0 \subset \Theta$ is *true* if $\theta_0 \in \Theta_0$.

In this section we discuss the *model selection problem*, that is, using the information provided by the sample in (5.1.1), we wish to find the "simplest" true model from a finite family of competing submodels associated with the sets

$$\Theta_1, \Theta_2, \ldots, \Theta_M \subseteq \Theta. \tag{5.1.2}$$

In the spirit of algebraic statistics, we assume the sets in (5.1.2) to be semi-algebraic (recall Definition 2.3.8). Moreover, as in previous sections we assume that the distributions P_{θ} have densities $p_{\theta}(x)$ with respect to some common dominating measure. In order to emphasize the role of the underlying observations, we denote the likelihood and log-likelihood function as

$$L_n(\theta \,|\, X^{(1)}, \ldots, X^{(n)}) \;=\; \prod_{i=1}^{n} p_\theta(X^{(i)}) \tag{5.1.3}$$

and $\ell_n(\theta \,|\, X^{(1)}, \ldots, X^{(n)}) = \log L_n(\theta \,|\, X^{(1)}, \ldots, X^{(n)})$, respectively.

Our approach to selecting true models among (5.1.2) is to search for models for which the maximized log-likelihood function

$$\hat{\ell}_n(i) \;=\; \sup_{\theta \in \Theta_i} \ell_n(\theta \,|\, X^{(1)}, \ldots, X^{(n)}) \tag{5.1.4}$$

is large. Of course, evaluating the quantity (5.1.4) requires solving the maximum likelihood estimation problem in Chapter 2. However, this methodology is not yet satisfactory since mere maximization of the log-likelihood function does not take into account differences in model complexity. In particular, $\Theta_1 \subset \Theta_2$ implies that $\hat{\ell}_n(1) \le \hat{\ell}_n(2)$. Information criteria provide a more refined approach.

Definition 5.1.1. The *information criterion* associated with a family of *penalty functions* $\pi_n \colon [M] \to \mathbb{R}$ assigns the score

$$\tau_n(i) = \hat{\ell}_n(i) - \pi_n(i)$$

to the i-th model, $i = 1, \ldots, M$.

The following are two classical examples of information criteria. Both measure model complexity in terms of dimension.

Example 5.1.2. The *Akaike information criterion* (AIC) due to [2] uses the penalty $\pi_n(i) = \dim(\Theta_i)$. The *Bayesian information criterion* (BIC) introduced in [82] uses the penalty $\pi_n(i) = \frac{\dim(\Theta_i)}{2} \log(n)$. $\qquad\square$

A score-based model search using an information criterion τ_n selects the model for which $\tau_n(i)$ is maximal. This approach has the consistency property in Theorem 5.1.3. This result is formulated in terms of regular exponential families. These were featured in Definition 2.3.11. As in Section 2.3, the details of the definition of this class of statistical models are not of importance here. It suffices to note that the class comprises very well-behaved models such as the family of all multivariate normal distributions and the interior of a probability simplex.

Theorem 5.1.3 (Consistency). *Consider a regular exponential family $\{P_\theta : \theta \in \Theta\}$. Let $\Theta_1, \Theta_2 \subseteq \Theta$ be arbitrary sets. Denote the ordinary closure of Θ_1 by $\overline{\Theta}_1$.*

(i) *Suppose $\theta_0 \in \Theta_2 \setminus \overline{\Theta}_1$. If the penalty functions are chosen such that the sequence $|\pi_n(2) - \pi_n(1)|/n$ converges to zero as $n \to \infty$, then*

$$P_{\theta_0}\left(\tau_n(1) < \tau_n(2)\right) \xrightarrow{n \to \infty} 1.$$

(ii) *Suppose $\theta_0 \in \Theta_1 \cap \Theta_2$. If the sequence of differences $\pi_n(1) - \pi_n(2)$ diverges to ∞ as $n \to \infty$, then*

$$P_{\theta_0}\left(\tau_n(1) < \tau_n(2)\right) \overset{n \to \infty}{\longrightarrow} 1.$$

For a proof of Theorem 5.1.3, see [56, Prop. 1.2]. Note that in [56] the result is stated for the case where Θ_1 and Θ_2 are smooth manifolds but this property is not used in the proof. In algebraic statistics, Θ_1 and Θ_2 will be semi-algebraic.

While the penalty for the AIC satisfies condition (i) but not (ii) in Theorem 5.1.3, it is straightforward to choose penalty functions that satisfy both (i) and (ii). For instance, the BIC penalty has this property. The original motivation for the BIC, however, is based on a connection to Bayesian model determination.

In the Bayesian approach we regard the data-generating distributions P_θ to be conditional distributions given the considered model being true and given the value of the parameter θ. In our setup, we thus assume that, given the i-th model and a parameter value $\theta \in \Theta_i$, the observations $X^{(1)}, \ldots, X^{(n)}$ are independent and identically distributed according to P_θ. We then express our (subjective) beliefs about the considered scenario by specifying a *prior distribution* over models and parameters. To this end, we choose a *prior* probability $P(\Theta_i)$, $i \in [M]$, for each of the competing models given by (5.1.2). And given the i-th model, we specify a (conditional) prior distribution Q_i for the parameter $\theta \in \Theta_i$.

After data are collected, statistical inference proceeds conditionally on the data. Being interested in model selection, we compute the *posterior probability* of the i-th model, namely, the conditional probability

$$P(\Theta_i \,|\, X^{(1)}, \ldots, X^{(n)}) \propto P(\Theta_i) \int_{\Theta_i} L_n(\theta \,|\, X^{(1)}, \ldots, X^{(n)}) dQ_i(\theta). \qquad (5.1.5)$$

Here we omitted the normalizing constant obtained by summing up the right hand sides for $i = 1, \ldots, M$. The difficulty in computing (5.1.5) is the evaluation of the integrated likelihood function, also known as *marginal likelihood integral*, or *marginal likelihood* for short. Integrals of this type are the topic of this chapter.

In typical applications, each set Θ_i is given parametrically as the image of some map $g_i \colon \mathbb{R}^d \to \mathbb{R}^k$, and the prior Q_i is specified via a distribution with Lebesgue density $p_i(\gamma)$ on $\mathbb{R}^d$. Suppressing the index i of the model, the marginal likelihood takes the form

$$\mu(X^{(1)}, \ldots, X^{(n)}) = \int_{\mathbb{R}^d} L_n\big(g(\gamma) \,|\, X^{(1)}, \ldots, X^{(n)}\big) p(\gamma) \, d\gamma \qquad (5.1.6a)$$

$$= \int_{\mathbb{R}^d} \exp\big\{\ell_n\big(g(\gamma) \,|\, X^{(1)}, \ldots, X^{(n)}\big)\big\} p(\gamma) \, d\gamma. \qquad (5.1.6b)$$

Example 5.1.4. Let $X^{(1)}, \ldots, X^{(n)}$ be independent $\mathcal{N}(\theta, Id_k)$ random vectors, $\theta \in \mathbb{R}^k$. In Example 2.3.2, we wrote the log-likelihood function of this model in terms of the sample mean $\bar{X}_n$. Plugging the expression into (5.1.6b) we see that the marginal likelihood is

$$\mu(X^{(1)}, \ldots, X^{(n)}) = \left(\frac{1}{\sqrt{(2\pi)^k}}\right)^n \exp\left\{-\frac{1}{2}\sum_{i=1}^n \|X^{(i)} - \bar{X}_n\|_2^2\right\}$$

$$\times \int_{\mathbb{R}^d} \exp\left\{-\frac{n}{2}\|\bar{X}_n - g(\gamma)\|_2^2\right\} p(\gamma)\, d\gamma. \quad (5.1.7)$$

Note that the factor

$$\left(\frac{1}{\sqrt{(2\pi)^k}}\right)^n \exp\left\{-\frac{1}{2}\sum_{i=1}^n \|X^{(i)} - \bar{X}_n\|_2^2\right\} \qquad (5.1.8)$$

is the maximized value of the likelihood function for $\theta \in \mathbb{R}^k$.   □

In Section 5.2, we discuss exact symbolic evaluation of marginal likelihood integrals in discrete models. In the present section, we will focus on the asymptotic behavior of integrals such as (5.1.7) when the sample size n is large. These allow one to approximate Bayesian model selection procedures. In particular, Theorem 5.1.5 below clarifies the connection between posterior model probabilities and the BIC.

For an asymptotic study, we shift back to the non-Bayesian setting of (5.1.1), in which we view the observations as drawn from some fixed unknown true distribution P_{θ_0}. In particular, we treat the marginal likelihood in (5.1.6a) as a sequence of random variables indexed by the sample size n and study its limiting behavior. Recall that a sequence of random variables (R_n) is *bounded in probability* if for all $\varepsilon > 0$ there exists a constant M_ε such that $P(|R_n| > M_\varepsilon) < \varepsilon$ for all n. We use the notation $O_p(1)$ for this property.

Theorem 5.1.5 (Laplace approximation). *Let $\{P_\theta : \theta \in \Theta\}$ be a regular exponential family with $\Theta \subseteq \mathbb{R}^k$. Consider an open set $\Gamma \subseteq \mathbb{R}^d$ and a smooth injective map $g : \Gamma \to \mathbb{R}^k$ that has continuous inverse on $g(\Gamma) \subseteq \Theta$. Let $\theta_0 = g(\gamma_0)$ be the true parameter, and assume that the Jacobian of g has full rank at γ_0 and that the prior density $p(\gamma)$ is a smooth function that is positive in a neighborhood of γ_0. Then*

$$\log \mu(X^{(1)}, \ldots, X^{(n)}) = \hat{\ell}_n - \frac{d}{2}\log(n) + O_p(1),$$

where

$$\hat{\ell}_n = \sup_{\gamma \in \Gamma} \ell_n(g(\gamma) \mid X^{(1)}, \ldots, X^{(n)}).$$

This theorem is proven in [56, Thm. 2.3], where a more refined expansion gives a remainder that is bounded in probability when multiplied by $\sqrt{n}$.

Theorem 5.1.5 shows that model selection using the BIC approximates a Bayesian procedure seeking the model with highest posterior probability. However, this approximation is only true for smooth models as we show in the next example.

Example 5.1.6 (Cuspidal cubic). Let $X^{(1)}, \ldots, X^{(n)}$ be independent $\mathcal{N}(\theta, Id_k)$ random vectors with $k = 2$. Following Example 2.3.6, we consider the cuspidal cubic model given by the parametrization $g(\gamma) = (\gamma^2, \gamma^3)$, $\gamma \in \mathbb{R}$. Let $\bar{X}_n = (\bar{X}_{n,1}, \bar{X}_{n,2})$ be the sample mean. By (5.1.7), evaluation of the marginal likelihood $\mu(X^{(1)}, \ldots, X^{(n)})$ requires the computation of the integral

$$\bar{\mu}(\bar{X}_n) = \int_{-\infty}^{\infty} \exp\left\{ -\frac{n}{2} \|\bar{X}_n - g(\gamma)\|_2^2 \right\} p(\gamma) \, d\gamma \tag{5.1.9}$$

$$= \int_{-\infty}^{\infty} \exp\left\{ -\frac{1}{2} [(\sqrt{n}\gamma^2 - \sqrt{n}\bar{X}_{n,1})^2 + (\sqrt{n}\gamma^3 - \sqrt{n}\bar{X}_{n,2})^3] \right\} p(\gamma) \, d\gamma.$$

If the true parameter $\theta_0 = g(\gamma_0)$ is non-zero, that is, $\gamma_0 \neq 0$, then Theorem 5.1.5 applies with $d = 1$. If, however, $\theta_0 = g(0) = 0$, then we find a different asymptotic behavior of the marginal likelihood. Changing variables to $\bar{\gamma} = n^{1/4}\gamma$ we obtain

$$\bar{\mu}(\bar{X}_n) = n^{-1/4} \int_{-\infty}^{\infty} \exp\left\{ -\frac{1}{2} \Big[(\bar{\gamma}^2 - \sqrt{n}\bar{X}_{n,1})^2 + \right.$$

$$\left. \left(\frac{\bar{\gamma}^3}{n^{1/4}} - \sqrt{n}\bar{X}_{n,2} \right)^3 \Big] \right\} p\left(\frac{\bar{\gamma}}{n^{1/4}} \right) d\bar{\gamma}. \tag{5.1.10}$$

By the central limit theorem, the independent sequences $\sqrt{n}\bar{X}_{n,1}$ and $\sqrt{n}\bar{X}_{n,2}$ each converge to the $\mathcal{N}(0, 1)$ distribution. Therefore, if Z_1 and Z_2 are independent $\mathcal{N}(0, 1)$ random variables, then

$$n^{1/4}\bar{\mu}(\bar{X}_n) \xrightarrow{D} \int_{-\infty}^{\infty} \exp\left\{ -\frac{1}{2} [(\gamma^2 - Z_1)^2 + Z_2^2] \right\} p(0) \, d\gamma. \tag{5.1.11}$$

Since convergence in distribution implies boundedness in probability, we obtain

$$\log \bar{\mu}(\bar{X}_n) = -\frac{1}{4} \log(n) + O_p(1).$$

It follows that

$$\log \mu(X^{(1)}, \ldots, X^{(n)}) = \hat{\ell}_n - \frac{1}{4} \log(n) + O_p(1) \tag{5.1.12}$$

if $\theta_0 = 0$. Note that $\hat{\ell}_n$ refers to the maximized log-likelihood function in the cuspidal cubic model. However, the difference between $\hat{\ell}_n$ and the factor in (5.1.8) is bounded in probability because it is equal to $1/2$ times a likelihood ratio statistic and thus converges in distribution according to Theorem 2.3.12. $\square$

The rates of convergence we computed in Example 5.1.6 have an interesting feature. If we replace the sample mean $\bar{X}_n$ in (5.1.9) by its expectation, which is the point $\theta_0 = g(\gamma_0)$ on the cuspidal cubic, then we obtain the integral

$$\bar{\mu}(\theta_0) = \int_{-\infty}^{\infty} \exp\left\{ -\frac{n}{2} \|g(\gamma_0) - g(\gamma)\|_2^2 \right\} p(\gamma) \, d\gamma. \tag{5.1.13}$$

This is a deterministic Laplace integral. If $\theta_0 \neq 0$, then the asymptotics of (5.1.13) can be determined using the classical Laplace approximation [100, p. 495] as follows:

$$\log \bar{\mu}(\theta_0) = -\frac{1}{2}\log(n) + O(1). \tag{5.1.14}$$

If $\theta_0 = 0$ then (5.1.10) implies

$$\log \bar{\mu}(\theta_0) = -\frac{1}{4}\log(n) + O(1). \tag{5.1.15}$$

This suggests that the asymptotic behavior of the marginal likelihood can be determined by studying the integral obtained by replacing the log-likelihood function in (5.1.6b) by its expectation under P_{θ_0}. This deterministic integral is equal to

$$\mu(\theta_0) = \int_{\mathbb{R}^d} \exp\{n\ell_0(g(\gamma))\}\, p(\gamma)\, d\gamma, \tag{5.1.16}$$

where $\ell_0(\theta) = \mathrm{E}[\log p_\theta(X)]$ for $X \sim P_{\theta_0}$. The function $\ell_0(\theta)$ for discrete models is obtained easily by writing $\log p_\theta(X)$ as in Example 2.3.10. In the Gaussian case, $\ell_0(\theta)$ can be found using that $\mathrm{E}[X] = \mu_0$ and $\mathrm{E}[XX^T] = \Sigma_0 + \mu_0\mu_0^T$, where μ_0 and Σ_0 are the mean vector and covariance matrix determined by θ_0.

The following result shows that what we observed in the example is true more generally.

Theorem 5.1.7. *Let $\{P_\theta : \theta \in \Theta\}$ be a regular exponential family with $\Theta \subseteq \mathbb{R}^k$. Consider an open set $\Gamma \subseteq \mathbb{R}^d$ and a polynomial map $g \colon \Gamma \to \Theta$. Let $\theta_0 = g(\gamma_0)$ be the true parameter. Assume that $g^{-1}(\theta_0)$ is a compact set and that the prior density $p(\gamma)$ is a smooth function on Γ that is positive on $g^{-1}(\theta_0)$. Then*

$$\log \mu(X^{(1)}, \dots, X^{(n)}) = \hat{\ell}_n - q\log(n) + (s-1)\log\log(n) + O_p(1),$$

where the rational number $q \in (0, d/2] \cap \mathbb{Q}$ and the integer $s \in [d]$ satisfy that

$$\log \mu(\theta_0) = n\ell_0(\theta_0) - q\log(n) + (s-1)\log\log(n) + O(1).$$

This and more general theorems are proven in the forthcoming book by Sumio Watanabe [97], which also gives an introduction to methods for computing the *learning coefficient* q and the *multiplicity* s in Theorem 5.1.7. These techniques are based on *resolution of singularities* in algebraic geometry. They have been applied to various mixture and hidden variable models; see e.g. [80, 101, 102, 103].

Example 5.1.8 (Reduced rank regression). Let $(X_1, \dots, X_m) \sim \mathcal{N}(0, \Sigma)$ be a multivariate normal random vector with mean zero, and consider a partition $A \cup B = [m]$ of the index set $[m]$. As mentioned in the proof of Proposition 3.1.13, the conditional distribution of X_A given $X_B = x_B$ is the multivariate normal distribution

$$\mathcal{N}\big(\Sigma_{A,B}\Sigma_{B,B}^{-1}x_B,\ \Sigma_{A,A} - \Sigma_{A,B}\Sigma_{B,B}^{-1}\Sigma_{B,A}\big).$$

Reduced rank regression is a Gaussian model in which the *matrix of regression coefficients* $\Sigma_{A,B}\Sigma_{B,B}^{-1}$ has low rank. In some instances, the requirement that the rank is at most h expresses the conditional independence of X_A and X_B given h hidden variables.

Let $a = \#A$ and $b = \#B$ and consider parametrizing $\Sigma_{A,B}\Sigma_{B,B}^{-1}$ as

$$g_h \colon \mathbb{R}^{a \times h} \times \mathbb{R}^{b \times h} \to \mathbb{R}^{a \times b},$$
$$(\alpha, \beta) \mapsto \alpha\beta^T.$$

The asymptotics of marginal likelihood integrals associated with this parametrization of reduced rank regression models were studied in a paper by Aoyagi and Watanabe [11]. The problem at the core of this work is to determine the asymptotics of integrals of the form

$$\int \exp\left\{-n\|\alpha\beta^T - \alpha_0\beta_0^T\|_2^2\right\} p(\alpha, \beta) \, d\alpha \, d\beta, \tag{5.1.17}$$

where $\alpha_0 \in \mathbb{R}^{a \times h}$ and $\beta_0 \in \mathbb{R}^{b \times h}$ are the true parameters. Since the preimages $g_h^{-1}(\theta_0)$ are not always compact, it is assumed in [11] that the prior density $p(\alpha, \beta)$ has compact support Ω and is positive at (α_0, β_0). Then the learning coefficient and associated multiplicity are derived for arbitrary values of a, b and h.

We illustrate here the case of rank $h = 1$. We assume that $a \geq b \geq 1$. If both $\alpha_0 \in \mathbb{R}^a$ and $\beta_0 \in \mathbb{R}^b$ are non-zero, then $g_1^{-1}(\alpha_0\beta_0^T) = \{(\alpha, \beta) : \alpha\beta^T = \alpha_0\beta_0^T\}$ is a one-dimensional smooth manifold, and the Jacobian of g_1 achieves its maximal rank $a + b - 1$ at (α_0, β_0). Therefore, in a neighborhood of $g_1(\alpha_0, \beta_0)$, the image of g_1 is an $a + b - 1$ dimensional smooth manifold. It follows from Theorem 5.1.5 that the learning coefficient is $q = (a + b - 1)/2$ and the multiplicity is $s = 1$.

The non-standard singular case occurs if $\alpha_0 = 0$ or $\beta_0 = 0$, in which case $g_1(\alpha_0, \beta_0) = 0$. As explained, for example, in the book by Watanabe [97] and his paper [96], the negated learning coefficient q is the largest pole of the zeta function. The *zeta function* is the meromorphic continuation of the function

$$\lambda \mapsto \int \left(\|\alpha\beta^T - \alpha_0\beta_0^T\|_2^2\right)^\lambda p(\alpha, \beta) \, d\alpha \, d\beta$$
$$= \int \left(\alpha_1^2 + \cdots + \alpha_a^2\right)^\lambda \left(\beta_1^2 + \cdots + \beta_b^2\right)^\lambda p(\alpha, \beta) \, d\alpha \, d\beta$$

from the set of complex numbers λ with $\mathrm{Re}(\lambda) > 0$ to the entire complex plane. The multiplicity s is the order of this pole.

Let $\Omega_\varepsilon = \Omega \cap \{(\alpha, \beta) : \|\alpha\beta^T\|_2^2 < \varepsilon\}$ for small $\varepsilon > 0$. Outside Ω_ε the integrand in (5.1.17) is bounded away from its maximum, and the asymptotic behavior of the integral remains unchanged if we restrict the integration domain to Ω_ε. We can cover Ω_ε by small neighborhoods $U(\alpha', \beta')$ around the singularities (α', β') with $\alpha' = 0$ or $\beta' = 0$. The learning coefficient q and the multiplicity s are determined by the most complicated singularity of g_1, which is at the origin $(\alpha', \beta') = 0$.

The resulting mathematical problem is to determine the poles of the integral

$$\int_{U(0,0)} \left(\alpha_1^2 + \cdots + \alpha_a^2\right)^\lambda \left(\beta_1^2 + \cdots + \beta_b^2\right)^\lambda p(\alpha, \beta)\, d\alpha\, d\beta. \tag{5.1.18}$$

This can be done using *blow-up* transformations. Here, we use the transformation

$$\alpha_1 = \alpha_1', \qquad \alpha_j = \alpha_1' \alpha_j' \quad \text{for all } j = 2, \ldots, a.$$

This map is a bijection for $\alpha_1 \neq 0$, and it transforms the integral in (5.1.18) to

$$\int_{U(0,0)} \alpha_1^{2\lambda} \left(1 + \alpha_2^2 + \cdots + \alpha_a^2\right)^\lambda \left(\beta_1^2 + \cdots + \beta_b^2\right)^\lambda \alpha_1^{a-1} p(\alpha, \beta)\, d\alpha\, d\beta,$$

where we have renamed the variables α_i' back to α_i. Using the analogous transformation for β we obtain the integral

$$\int_{U(0,0)} \alpha_1^{2\lambda} \beta_1^{2\lambda} \left(1 + \alpha_2^2 + \cdots + \alpha_a^2\right)^\lambda \left(1 + \beta_2^2 + \cdots + \beta_b^2\right)^\lambda \alpha_1^{a-1} \beta_1^{b-1} p(\alpha, \beta)\, d\alpha\, d\beta.$$

The product $\left(1 + \alpha_2^2 + \cdots + \alpha_a^2\right)\left(1 + \beta_2^2 + \cdots + \beta_b^2\right)$ and the prior density $p(\alpha, \beta)$ are bounded away from 0 in the neighborhood $U(0,0)$. Therefore, we may consider

$$\int \alpha_1^{2\lambda+a-1} \beta_1^{2\lambda+b-1}\, d\alpha_1 d\beta_1 \;=\; \frac{\alpha_1^{2\lambda+a} \beta_1^{2\lambda+b}}{(2\lambda + a)(2\lambda + b)}.$$

As a function of λ, this integral has poles at $\lambda = -a/2$ and $\lambda = -b/2$. Having assumed that $a \geq b$, the larger pole is $-b/2$, and thus the learning coefficient is $q = b/2$. The multiplicity is $s = 2$ if $a = b$, and it is $s = 1$ if $a > b$. □

Blow-up transformations are the engine behind algorithms for resolutions of singularities in algebraic geometry. An implementation of such a resolution algorithm can be found in `Singular`. In theory, this implementation can be used to obtain information about the asymptotic behavior of Laplace integrals. However, in practice, we found it prohibitive to use a general algorithm for resolution of singularities, because of the enormous complexity in running time and output size. On the other hand, polyhedral geometry and the theory of toric varieties furnish combinatorial tools for resolution of singularities. These work well under suitable genericity assumptions. We conclude this section by showing an example.

Example 5.1.9 (Remoteness and Laplace integrals). Let l and k be two even positive integers and consider the integral

$$\mu_{k,l} \;=\; \int_{-\infty}^{\infty} \int_{-\infty}^{\infty} e^{-n(x^k + y^l)}\, dx\, dy,$$

which is a product of two univariate integrals that can be computed in closed form, e.g. using `Maple`. We find that the logarithm of the integral equals

$$\log \mu_{k,l} \;=\; -\left(\frac{1}{k} + \frac{1}{l}\right) \log(n) + O(1). \tag{5.1.19}$$

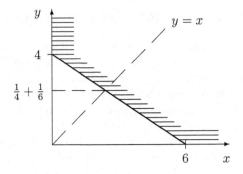

Figure 5.1.1: Newton diagram for $x^6 + y^4$.

The coefficient $q = 1/k + 1/l$ can be obtained from the *Newton diagram* of the phase function $x^k + y^l$. The Newton diagram is the convex hull of the set obtained by attaching the non-negative orthant $[0, \infty)^2$ to each of the exponents $(k, 0)$ and $(0, l)$ appearing in the phase function; see Figure 5.1.1. The *remoteness* of the Newton diagram is the reciprocal $1/\rho$ of the smallest positive real ρ such that $\rho \cdot (1, 1)$ is in the Newton diagram. In this example, $\rho = kl/(k + l)$, and the remoteness is found to be $1/\rho = 1/k + 1/l$. This coincides with the coefficient q appearing in the integral (5.1.19). □

In general, the phase function needs to exhibit certain non-degenerateness conditions for the remoteness of the Newton diagram to be equal to the learning coefficient q. If the conditions apply, then the codimension of the face in which the diagonal spanned by the vector $(1, \ldots, 1)^T$ first hits the Newton diagram determines the multiplicity s for the $\log \log n$ term. We illustrate this in the reduced rank regression example below and refer the reader to the book [12, §8.3.2] for the precise results. In many other statistical models, however, the required non-degeneracy conditions do not apply. Extending the scope of Newton diagram methods is thus an important topic for future work.

Example 5.1.10 (Remoteness in reduced rank regression). Suppose $a = b = 2$ in the reduced rank regression problem considered in Example 5.1.8. The Newton diagram of the phase function $(\alpha_1^2 + \alpha_2^2)(\beta_1^2 + \beta_2^2)$ in (5.1.18) has the vertices $(2, 0, 2, 0)$, $(2, 0, 0, 2)$, $(0, 2, 2, 0)$ and $(0, 2, 0, 2)$. It can also be described by the inequalities

$$\alpha_1, \alpha_2, \beta_1, \beta_2 \geq 0, \quad \alpha_1 + \alpha_2 \geq 2, \quad \beta_1 + \beta_2 \geq 2.$$

If $\rho \cdot (1, 1, 1, 1)$ is in the Newton diagram then $2\rho \geq 2$. The minimum feasible value for ρ is thus 1, and we find that the remoteness, being also equal to 1, gives the correct learning coefficient $b/2 = 2/2 = 1$. Since the point $(1, 1, 1, 1)$ lies on a two-dimensional face of the Newton diagram, the multiplicity is $s = 2$. □

5.2   Exact Integration for Discrete Models

Inference in Bayesian statistics involves the evaluation of marginal likelihood integrals. We present algebraic methods for computing such integrals exactly for discrete data of small sample size. The relevant models are mixtures of independent probability distributions, or, in geometric language, secant varieties of Segre-Veronese varieties. Our approach applies to both uniform priors and Dirichlet priors. This section is based on the paper [69]. Our notational conventions in this section differ slightly from the rest of this book. These differences make for simpler formulas and allow for easier use of the Maple code described later.

We consider a collection of discrete random variables

$$
\begin{array}{cccc}
X_1^{(1)}, & X_2^{(1)}, & \ldots, & X_{s_1}^{(1)}, \\
X_1^{(2)}, & X_2^{(2)}, & \ldots, & X_{s_2}^{(2)}, \\
\vdots & \vdots & \ddots & \vdots \\
X_1^{(m)}, & X_2^{(m)}, & \ldots, & X_{s_m}^{(m)},
\end{array}
$$

where the variables in the i-th row, $X_1^{(i)}, X_2^{(i)}, \ldots, X_{s_i}^{(i)}$, are identically distributed with values in $\{0, 1, \ldots, r_i\}$. The independence model $\mathcal{M}$ for these variables is a log-linear model as in Chapter 1. It is represented by a $d \times k$-matrix A with

$$
d = r_1 + r_2 + \cdots + r_m + m \quad \text{and} \quad k = \prod_{i=1}^{m}(r_i + 1)^{s_i}. \tag{5.2.1}
$$

The columns of the matrix A are indexed by elements v of the state space

$$
\{0, 1, \ldots, r_1\}^{s_1} \times \{0, 1, \ldots, r_2\}^{s_2} \times \cdots \times \{0, 1, \ldots, r_m\}^{s_m}. \tag{5.2.2}
$$

The rows of the matrix A are indexed by the model parameters, which are the d coordinates of the points $\theta = (\theta^{(1)}, \theta^{(2)}, \ldots, \theta^{(m)})$ in the product of simplices

$$
\Theta = \Delta_{r_1} \times \Delta_{r_2} \times \cdots \times \Delta_{r_m}. \tag{5.2.3}
$$

The *independence model* $\mathcal{M}$ is the subset of the probability simplex Δ_{k-1} which is given by the parametrization

$$
p_v = P\big(X_j^{(i)} = v_j^{(i)} \text{ for all } i, j\big) = \prod_{i=1}^{m}\prod_{j=1}^{s_i} \theta_{v_j^{(i)}}^{(i)}. \tag{5.2.4}
$$

This expression is a monomial in d unknowns. The column vector a_v of the matrix A corresponding to state v is the exponent vector of the monomial in (5.2.4).

For an algebraic geometer, the model $\mathcal{M}$ is the *Segre-Veronese variety*

$$
\mathbb{P}^{r_1} \times \mathbb{P}^{r_2} \times \cdots \times \mathbb{P}^{r_m} \hookrightarrow \mathbb{P}^{k-1}, \tag{5.2.5}
$$

where the embedding is given by the line bundle $\mathcal{O}(s_1, s_2, \ldots, s_m)$. The manifold $\mathcal{M}$ is the toric variety of the polytope Θ. Both objects have dimension $d - m$.

Example 5.2.1. Consider three binary random variables where the last two random variables are identically distributed. In our notation, this corresponds to $m = 2$, $s_1 = 1$, $s_2 = 2$ and $r_1 = r_2 = 1$. We find that $d = 4, k = 8$, and

$$
A \;=\; \begin{array}{c} \\ \theta_0^{(1)} \\ \theta_1^{(1)} \\ \theta_0^{(2)} \\ \theta_1^{(2)} \end{array}
\begin{array}{c} p_{000}\ \ p_{001}\ \ p_{010}\ \ p_{011}\ \ p_{100}\ \ p_{101}\ \ p_{110}\ \ p_{111} \\
\left(\begin{array}{cccccccc}
1 & 1 & 1 & 1 & 0 & 0 & 0 & 0 \\
0 & 0 & 0 & 0 & 1 & 1 & 1 & 1 \\
2 & 1 & 1 & 0 & 2 & 1 & 1 & 0 \\
0 & 1 & 1 & 2 & 0 & 1 & 1 & 2
\end{array}\right). \end{array}
$$

The columns of this matrix represent the monomials in the parametrization (5.2.4). The model $\mathcal{M}$ is a surface known to geometers as a *rational normal scroll*. Its Markov basis consists of the 2×2-minors of the matrix

$$
\begin{pmatrix} p_{000} & p_{001} & p_{100} & p_{101} \\ p_{010} & p_{011} & p_{110} & p_{111} \end{pmatrix}
$$

together with the two linear relations $p_{001} = p_{010}$ and $p_{101} = p_{110}$. □

The matrix A has repeated columns whenever $s_i \geq 2$ for some i. It is usually preferable to represent the model $\mathcal{M}$ by the matrix $\tilde{A}$ which is obtained from A by removing repeated columns. We label the columns of the matrix $\tilde{A}$ by elements $v = (v^{(1)}, \ldots, v^{(m)})$ of (5.2.2) whose components $v^{(i)} \in \{0, 1, \ldots, r_i\}^{s_i}$ are weakly increasing. Hence $\tilde{A}$ is a $d \times \tilde{k}$-matrix with

$$
\tilde{k} \;=\; \prod_{i=1}^{m} \binom{s_i + r_i}{s_i}. \tag{5.2.6}
$$

The model $\mathcal{M}$ and its mixtures are subsets of a subsimplex $\Delta_{\tilde{k}-1}$ of Δ_{k-1}.

We now examine Bayesian integrals for the independence model $\mathcal{M}$. All domains of integration in this section are products of standard probability simplices. On each such polytope we fix the standard Lebesgue probability measure. This corresponds to taking uniform priors $p(\gamma) = 1$ in the integral (5.1.6a). Naturally, other prior distributions, such as the conjugate Dirichlet priors, are of interest, and our methods will be extended to these in Corollary 5.2.8. For now, we simply fix uniform priors. We identify the state space (5.2.2) with the set $\{1, \ldots, k\}$. A data vector $u = (u_1, \ldots, u_k)$ is thus an element of $\mathbb{N}^k$. The sample size of these data is $\|u\|_1 = n$. The likelihood function (5.1.3) for these data equals

$$
L(\theta) \;=\; \frac{n!}{u_1! u_2! \cdots u_k!} \cdot p_1(\theta)^{u_1} \cdot p_2(\theta)^{u_2} \cdots \cdot p_k(\theta)^{u_k}.
$$

This expression is a polynomial function on the polytope Θ in (5.2.3). The marginal likelihood (5.1.6a) for the independence model $\mathcal{M}$ equals

$$
\int_{\Theta} L(\theta)\, d\theta.
$$

The value of this integral is a rational number that we now compute explicitly. The data u enter this calculation by way of the *sufficient statistics* $b = Au$. The vector b is in $\mathbb{N}^d$, and its coordinates $b_j^{(i)}$ for $i = 1, \ldots, m$ and $j = 0, \ldots, r_m$ count the total number of times each value j is attained by one of the random variables $X_1^{(i)}, \ldots, X_{s_i}^{(i)}$ in the i-th row. The sufficient statistics satisfy

$$b_0^{(i)} + b_1^{(i)} + \cdots + b_{r_i}^{(i)} \;=\; s_i \cdot n \qquad \text{for all } i = 1, 2, \ldots, m. \tag{5.2.7}$$

Our likelihood function $L(\theta)$ is the constant $n!/(u_1! \cdots u_k!)$ times the monomial

$$\theta^b \;=\; \prod_{i=1}^{m} \prod_{j=0}^{r_i} (\theta_j^{(i)})^{b_j^{(i)}}.$$

We note that the independence model $\mathcal{M}$ has maximum likelihood degree 1:

Remark 5.2.2. *The function θ^b is concave on the polytope Θ, and its maximum value is attained at the point $\hat{\theta}$ with coordinates $\hat{\theta}_j^{(i)} = b_j^{(i)}/(s_i \cdot n)$.*

Not just maximum likelihood estimation but also Bayesian integration is very easy for the model $\mathcal{M}$:

Lemma 5.2.3. *The marginal likelihood integral for the independence model equals*

$$\int_{\Theta} L(\theta) \, d\theta \;=\; \frac{n!}{u_1! \cdots u_k!} \cdot \prod_{i=1}^{m} \frac{r_i! \, b_0^{(i)}! \, b_1^{(i)}! \cdots b_{r_i}^{(i)}!}{(s_i n + r_i)!}.$$

Proof. Since Θ is the product of simplices (5.2.3), this follows from the formula

$$\int_{\Delta_t} \theta_0^{b_0} \theta_1^{b_1} \cdots \theta_t^{b_t} \, d\theta \;=\; \frac{t! \cdot b_0! \cdot b_1! \cdots b_t!}{(b_0 + b_1 + \cdots + b_t + t)!} \tag{5.2.8}$$

for the integral of a monomial over the standard probability simplex Δ_t. $\quad\square$

We now come to our main objective, which is to compute marginal likelihood integrals for the mixture model $\mathrm{Mixt}^2(\mathcal{M})$. Our parameter space is the polytope

$$\Theta^{(2)} \;=\; \Delta_1 \times \Theta \times \Theta.$$

The mixture model $\mathrm{Mixt}^2(\mathcal{M})$ is the subset of Δ_{k-1} with the parametric representation

$$p_v \;=\; \sigma_0 \cdot \theta^{a_v} + \sigma_1 \cdot \rho^{a_v} \qquad \text{for } (\sigma, \theta, \rho) \in \Theta^{(2)}. \tag{5.2.9}$$

Here $a_v \in \mathbb{N}^d$ is the column vector of A indexed by the state v, which is either in (5.2.2) or in $\{1, 2, \ldots, k\}$. The likelihood function of the mixture model equals

$$L(\sigma, \theta, \rho) \;=\; \frac{n!}{u_1! u_2! \cdots u_k!} \, p_1(\sigma, \theta, \rho)^{u_1} \cdots p_k(\sigma, \theta, \rho)^{u_k}, \tag{5.2.10}$$

and the marginal likelihood for the model $\text{Mixt}^2(\mathcal{M})$ equals

$$\int_{\Theta^{(2)}} L(\sigma, \theta, \rho) \, d\sigma d\theta d\rho \;=\; \frac{n!}{u_1! \cdots u_k!} \int_{\Theta^{(2)}} \prod_v (\sigma_0 \theta^{a_v} + \sigma_1 \rho^{a_v})^{u_v} d\sigma \, d\theta \, d\rho. \quad (5.2.11)$$

Proposition 5.2.4. *The marginal likelihood* (5.2.11) *is a rational number.*

Proof. The likelihood function $L(\sigma, \theta, \rho)$ is a $\mathbb{Q}_{\geq 0}$-linear combination of monomials $\sigma^a \theta^b \rho^c$. The integral (5.2.11) is the same $\mathbb{Q}_{\geq 0}$-linear combination of the numbers

$$\int_{\Theta^{(2)}} \sigma^a \theta^b \rho^c \, d\sigma \, d\theta \, d\rho \;=\; \int_{\Delta_1} \sigma^a \, d\sigma \cdot \int_\Theta \theta^b \, d\theta \cdot \int_\Theta \rho^c \, d\rho.$$

Each of the three factors is an easy-to-evaluate rational number, by (5.2.8). $\qquad\square$

The model $\text{Mixt}^2(\mathcal{M})$ corresponds to the first secant variety (see Section 4.1) of the Segre-Veronese variety (5.2.5). We could also consider the higher mixture models $\text{Mixt}^l(\mathcal{M})$, which correspond to mixtures of l independent distributions, and much of our analysis can be extended to that case, but for simplicity we restrict ourselves to $l = 2$. The secant variety $\text{Sec}^2(\overline{\mathcal{M}})$ is embedded in the projective space $\mathbb{P}^{\tilde{k}-1}$ with $\tilde{k}$ as in (5.2.6). Note that $\tilde{k}$ can be much smaller than k. If this is the case then it is convenient to aggregate states whose probabilities are identical and to represent the data by a vector $\tilde{u} \in \mathbb{N}^{\tilde{k}}$. Here is an example.

Example 5.2.5. Let $m=1$, $s_1=4$ and $r_1=1$, so $\mathcal{M}$ is the independence model for four identically distributed binary random variables. Then $d = 2$ and $k = 16$. The corresponding integer matrix and its row and column labels are

$$A \;=\; \begin{matrix} \\ \theta_0 \\ \theta_1 \end{matrix} \begin{matrix} p_{0000} & p_{0001} & p_{0010} & p_{0100} & p_{1000} & p_{0011} & \cdots & p_{1110} & p_{1111} \\ \left(\begin{matrix} 4 & 3 & 3 & 3 & 3 & 2 & \cdots & 1 & 0 \\ 0 & 1 & 1 & 1 & 1 & 2 & \cdots & 3 & 4 \end{matrix} \right. & & & & & & & & \left. \vphantom{\begin{matrix}4\\0\end{matrix}} \right). \end{matrix}$$

However, this matrix has only $\tilde{k} = 5$ distinct columns, and we instead use

$$\tilde{A} \;=\; \begin{matrix} \\ \theta_0 \\ \theta_1 \end{matrix} \begin{matrix} p_0 & p_1 & p_2 & p_3 & p_4 \\ \left(\begin{matrix} 4 & 3 & 2 & 1 & 0 \\ 0 & 1 & 2 & 3 & 4 \end{matrix} \right). \end{matrix}$$

The mixture model $\text{Mixt}^2(\mathcal{M})$ is a Zariski dense subset of the cubic hypersurface in Δ_4 that was discussed in Example 2.2.3, where we studied the likelihood function (2.2.2) for the data vector

$$\tilde{u} \;=\; (\tilde{u}_0, \tilde{u}_1, \tilde{u}_2, \tilde{u}_3, \tilde{u}_4) \;=\; (51, 18, 73, 25, 75).$$

It has three local maxima (modulo swapping θ and ρ) whose coordinates are algebraic numbers of degree 12. Using the `Maple` library cited below, we computed the exact value of the marginal likelihood for the data $\tilde{u}$. The number (5.2.11) is the ratio of two relatively prime integers having 530 digits and 552 digits, and its numerical value is approximately $0.7788716338838678611335742860090 \cdot 10^{-22}$. $\qquad\square$

Algorithms for computing (5.2.11) are sketched below, and described in detail in [69]. These algorithms are implemented in a `Maple` library which is available at

$$\text{http://math.berkeley.edu/}\sim\text{shaowei/integrals.html.}$$

The input for that `Maple` code consists of parameter vectors $s = (s_1, \ldots, s_m)$ and $r = (r_1, \ldots, r_m)$ as well as a data vector $u \in \mathbb{N}^k$. This input uniquely specifies the $d \times k$-matrix A. Here d and k are as in (5.2.1). Output features include the exact rational values of the marginal likelihood for both $\mathcal{M}$ and $\text{Mixt}^2(\mathcal{M})$.

Example 5.2.6. ("Schizophrenic patients") We apply our exact integration method to a data set taken from the Bayesian statistics literature. Evans, Gilula and Guttman [45, §3] analyzed the association between length of hospital stay (in years Y) of 132 schizophrenic patients and the frequency with which they are visited by their relatives. The data vector u for their data set is determined by the 3×3-contingency table:

	$2 \leq Y < 10$	$10 \leq Y < 20$	$20 \leq Y$	Totals
Visited regularly	43	16	3	62
Visited rarely	6	11	10	27
Visited never	9	18	16	43
Totals	58	45	29	132

The paper [45] presents estimated posterior means and variances for these data, where *"each estimate requires a 9-dimensional integration"* [45, p. 561]. Computing these integrals is essentially equivalent to our integrals, for $m = 2, s_1 = s_2 = 1, r_1 = r_2 = 2$ and $n = 132$. The authors point out that *"all posterior moments can be calculated in closed form .... however, even for modest n these expressions are far too complicated to be useful"* [45, p. 559] and emphasize that *"the dimensionality of the integral does present a problem"* [45, p. 562].

We disagree with the conclusion that closed form expressions are not useful. In our view, exact integration is quite practical for modest sample size such as $n = 132$. Using the `Maple` library, we computed the integral in (5.2.11). It is the rational number with numerator

> 2780194885310633891206436003249893291038761408055
> 2852428395820925693572658866753228458740975280339
> 9949306971310363319990693940571118083756885373737

and denominator

> 12288402873591935400678094796599848745442833177572204
> 50448819979286456995185542195946815073112429169997801
> 33503900169921912167352239204153786645029153951176422
> 43298328046163472261962028461650432024356339706541132
> 343753184718802748186676574237491200000000000000000.

To obtain the marginal likelihood for the data u above, the integral has to be multiplied with the normalizing constant $132!/(43!\cdot16!\cdot3!\cdot6!\cdot11!\cdot10!\cdot9!\cdot18!\cdot16!)$. ☐

Our approach to exact integration makes use of the fact that the Newton polytope of the likelihood function (5.2.10) is a zonotope. Recall that the *Newton polytope* of a polynomial is the convex hull of all exponent vectors appearing in the expansion of that polynomial, and a polytope is a *zonotope* if it is the image of a standard cube under a linear map; see Chapter 7 of [105] for more on zonotopes. We are here considering the zonotope

$$Z_A(u) = \sum_{v=1}^{k} u_v \cdot [0, a_v],$$

where $[0, a_v]$ is the line segment between the origin and the point $a_v \in \mathbb{R}^d$. So, the zonotope $Z_A(u)$ is the Minkowski sum of these k segments. Assuming that the counts u_v are all positive, we have

$$\dim(Z_A(u)) = \operatorname{rank}(A) = d - m + 1. \tag{5.2.12}$$

The zonotope $Z_A(u)$ is related to the polytope $\Theta = \operatorname{conv}(A)$ in (5.2.3) as follows. The dimension $d - m = r_1 + \cdots + r_m$ of Θ is one less than $\dim(Z_A(u))$, and Θ appears as the *vertex figure* of the zonotope $Z_A(u)$ at the distinguished vertex 0.

The marginal likelihood to be computed is $n!/(u_1! \cdots u_k!)$ times the integral

$$\int_{\Theta^{(2)}} \prod_{v=1}^{k} (\sigma_0 \theta^{a_v} + \sigma_1 \rho^{a_v})^{u_v} \, d\sigma d\theta d\rho. \tag{5.2.13}$$

Our approach to this computation is to sum over the lattice points in the zonotope $Z_A(u)$. If the matrix A has repeated columns, we may replace A with the reduced matrix $\tilde{A}$ and u with the corresponding reduced data vector $\tilde{u}$. This would require introducing an extra normalizing constant. In what follows we simply focus on computing the integral (5.2.13) with respect to the original matrix A. Recall from (5.2.4) that all columns of the matrix A have the same coordinate sum

$$a := |a_v| = s_1 + s_2 + \cdots + s_m \quad \text{for all } v = 1, 2, \ldots, k,$$

and from (5.2.7) that we may denote the entries of a vector $b \in \mathbb{R}^d$ by $b_j^{(i)}$ for $i = 1, \ldots, m$ and $j = 0, \ldots, r_m$. Also, let $\mathcal{L}$ denote the image of the linear map $A: \mathbb{Z}^k \to \mathbb{Z}^d$. Thus $\mathcal{L}$ is a sublattice of rank $d - m + 1$ in $\mathbb{Z}^d$. We abbreviate $Z_A^{\mathcal{L}}(u) := Z_A(u) \cap \mathcal{L}$. Expanding the integrand in (5.2.13) gives

$$\prod_v (\sigma_0 \theta^{a_v} + \sigma_1 \rho^{a_v})^{u_v} = \sum_{\substack{b \in Z_A^{\mathcal{L}}(u) \\ c = Au - b}} \phi_A(b, u) \cdot \sigma_0^{|b|/a} \sigma_1^{|c|/a} \cdot \theta^b \cdot \rho^c. \tag{5.2.14}$$

Writing $\mathcal{D}(u) = \{(x_1, \ldots, x_k) \in \mathbb{Z}^k : 0 \le x_v \le u_v, v = 1, \ldots, k\}$, we can see that the coefficient in (5.2.14) equals

$$\phi_A(b, u) = \sum_{\substack{Ax=b \\ x \in \mathcal{D}(u)}} \prod_{v=1}^{k} \binom{u_v}{x_v}. \tag{5.2.15}$$

Thus, by formulas (5.2.8) and (5.2.14), the integral (5.2.13) evaluates to

$$\sum_{\substack{b \in Z_A^c(u) \\ c = Au - b}} \phi_A(b, u) \cdot \frac{(|b|/a)! \, (|c|/a)!}{(|u| + 1)!} \cdot \prod_{i=1}^{m} \left(\frac{r_i! \, b_0^{(i)}! \cdots b_{r_i}^{(i)}!}{(|b^{(i)}| + r_i)!} \; \frac{r_i! \, c_0^{(i)}! \cdots c_{r_i}^{(i)}!}{(|c^{(i)}| + r_i)!} \right). \quad (5.2.16)$$

We summarize the result of this derivation in the following theorem.

Theorem 5.2.7. *The marginal likelihood for the mixture model* $\mathrm{Mixt}^2(\mathcal{M})$ *is equal to the sum* (5.2.16) *times the normalizing constant* $n!/(u_1! \cdots u_k!)$.

Each individual summand in the formula (5.2.16) is a ratio of factorials and hence can be evaluated symbolically. The challenge in turning Theorem 5.2.7 into a practical algorithm lies in the fact that both of the sums (5.2.15) and (5.2.16) are over very large sets. These challenges are addressed by the methods in [69, §4].

We now remove the restrictive assumption that our marginal likelihood integral has been evaluated with respect to the uniform distribution (Lebesgue measure) on the parameter space $\Theta^{(2)}$. In Bayesian analysis of discrete data, it is standard practice to compute such integrals with respect to the *Dirichlet prior distributions*, which form conjugate priors to multinomial models like our independence model $\mathcal{M}$. We shall show in Corollary 5.2.8 how the formula in Theorem 5.2.7 extends from uniform priors to Dirichlet priors.

The *Dirichlet distribution* $\mathrm{Dir}(\alpha)$ is a continuous probability distribution that is parametrized by a vector $\alpha = (\alpha_0, \alpha_1, \ldots, \alpha_t)$ of positive reals. The probability density function $f(\theta; \alpha)$ of $\mathrm{Dir}(\alpha)$ is supported on the t-dimensional simplex

$$\Delta_t = \{ (\theta_0, \ldots, \theta_t) \in \mathbb{R}_{\geq 0}^t : \theta_0 + \cdots + \theta_t = 1 \},$$

and it equals

$$f(\theta_0, \ldots, \theta_t; \alpha_0, \ldots, \alpha_t) = \frac{1}{B(\alpha)} \cdot \theta_0^{\alpha_0 - 1} \theta_1^{\alpha_1 - 1} \cdots \theta_t^{\alpha_t - 1} =: \frac{\theta^{\alpha - 1}}{B(\alpha)}.$$

Here the normalizing constant is the multivariate beta function

$$B(\alpha) = \frac{t! \cdot \Gamma(\alpha_0) \cdot \Gamma(\alpha_1) \cdots \Gamma(\alpha_t)}{\Gamma(\alpha_0 + \alpha_1 + \cdots + \alpha_t)}.$$

Note that, if the α_i are all integers, then this is the rational number

$$B(\alpha) = \frac{t! \cdot (\alpha_0 - 1)! \cdot (\alpha_1 - 1)! \cdots (\alpha_t - 1)!}{(\alpha_0 + \cdots + \alpha_t - 1)!}.$$

Thus the identity (5.2.8) is the special case of the identity $\int_{\Delta_t} f(\theta; \alpha) \, d\theta = 1$ for the density of the Dirichlet distribution when all $\alpha_i = b_i + 1$ are integers.

We now show how to compute marginal likelihood integrals for the mixture model $\mathrm{Mixt}^2(\mathcal{M})$ with respect to Dirichlet priors. Fix positive vectors $\alpha \in (0, \infty)^2$, and $\beta^{(i)}, \gamma^{(i)} \in (0, \infty)^{r_i+1}$ for $i = 1, \ldots, m$. These determine the distribution $\mathrm{Dir}(\alpha)$ on Δ_1, and the distributions $\mathrm{Dir}(\beta^{(i)})$ and $\mathrm{Dir}(\gamma^{(i)})$ on the i-th factor Δ_{r_i} in the product (5.2.3). The product probability measure

$$\mathrm{Dir}(\alpha) \otimes \prod_{i=1}^{m} \mathrm{Dir}(\beta^{(i)}) \otimes \prod_{i=1}^{m} \mathrm{Dir}(\gamma^{(i)})$$

is a distribution on $\Theta^{(2)} = \Delta_1 \times \Theta \times \Theta$ that we call the *Dirichlet distribution* with parameters (α, β, γ). Its probability density function is the product of the respective densities:

$$f(\sigma, \theta, \rho; \alpha, \beta, \gamma) = \frac{\sigma^{\alpha-1}}{B(\alpha)} \cdot \prod_{i=1}^{m} \frac{(\theta^{(i)})^{\beta^{(i)}-1}}{B(\beta^{(i)})} \cdot \prod_{i=1}^{m} \frac{(\rho^{(i)})^{\gamma^{(i)}-1}}{B(\gamma^{(i)})}. \qquad (5.2.17)$$

By the marginal likelihood with Dirichlet priors we mean the integral

$$\int_{\Theta^{(2)}} L(\sigma, \theta, \rho) \, f(\sigma, \theta, \rho; \alpha, \beta, \gamma) \, d\sigma d\theta d\rho. \qquad (5.2.18)$$

This is a modification of (5.2.11), which depends not just on the data u and the model $\mathrm{Mixt}^2(\mathcal{M})$ but also on the choice of Dirichlet parameters (α, β, γ). When the coordinates of these parameters are arbitrary positive reals but not integers, then the value of the integral (5.2.18) is no longer a rational number. Nonetheless, it can be computed exactly as follows. We abbreviate the product of Gamma functions in the denominator of the density (5.2.17) as follows:

$$B(\alpha, \beta, \gamma) := B(\alpha) \cdot \prod_{i=1}^{m} B(\beta^{(i)}) \cdot \prod_{i=1}^{m} B(\gamma^{(i)}).$$

Instead of the integrand (5.2.14) we now need to integrate

$$\sum_{\substack{b \in Z_A^{\mathcal{L}}(u) \\ c=Au-b}} \frac{\phi_A(b, u)}{B(\alpha, \beta, \gamma)} \cdot \sigma_0^{|b|/a+\alpha_0-1} \cdot \sigma_1^{|c|/a+\alpha_1-1} \cdot \theta^{b+\beta-1} \cdot \rho^{c+\gamma-1}$$

with respect to Lebesgue probability measure on $\Theta^{(2)}$. Doing this term by term, as before, we obtain the following modification of Theorem 5.2.7.

Corollary 5.2.8. *The marginal likelihood for the mixture model $\mathrm{Mixt}^2(\mathcal{M})$ with respect to Dirichlet priors with parameters (α, β, γ) equals*

$$\frac{n!}{u_1! \cdots u_k! \cdot B(\alpha, \beta, \gamma)} \sum_{\substack{b \in Z_A^{\mathcal{L}}(u) \\ c=Au-b}} \phi_A(b, u) \frac{\Gamma(|b|/a + \alpha_0)\Gamma(|c|/a + \alpha_1)}{\Gamma(|u| + |\alpha|)}$$

$$\times \prod_{i=1}^{m} \left(\frac{r_i! \, \Gamma(b_0^{(i)} + \beta_0^{(i)}) \cdots \Gamma(b_{r_i}^{(i)} + \beta_{r_i}^{(i)})}{\Gamma(|b^{(i)}| + |\beta^{(i)}|)} \frac{r_i! \, \Gamma(c_0^{(i)} + \gamma_0^{(i)}) \cdots \Gamma(c_{r_i}^{(i)} + \gamma_{r_i}^{(i)})}{\Gamma(|c^{(i)}| + |\gamma^{(i)}|)} \right).$$

Chapter 6

Exercises

This chapter presents solutions to eight problems. Problems 6.1 and 6.2 are concerned with Markov bases and can be solved based on the material presented in Chapter 1 alone. Problem 6.3 is about a Gaussian graphical model and relies on Section 2.1. Problems 6.4 and 6.5 complement Section 2.3 by providing a detailed study of the asymptotics of the likelihood ratio test in a Gaussian model and a calculation of a Fisher-information matrix. Problems 6.6 and 6.7 illustrate the use of algebraic techniques to study implications among conditional independence relations as discussed in Chapter 3. Problem 6.8 concerns a mixture model (recall Section 4.1) and involves in particular questions about exact computation of Bayesian integrals (recall Section 5.2). For each problem we list the names of the team members who worked on this problem during the week in Oberwolfach. These teams submitted their solutions to Florian Block, Dustin Cartwright, Filip Cools and Alex Engström, who helped compile them into the current chapter.

6.1 Markov Bases Fixing Subtable Sums

The team consisted of Jörn Dannemann, Hajo Holzmann and Or Zuk.

Problem. Let $S \subset [r] \times [c]$. Consider the log-linear model for a 2-way contingency table given parametrically by

$$\log p_{ij} = \begin{cases} \alpha_i + \beta_j + \lambda & \text{if } (i,j) \in S, \\ \alpha_i + \beta_j & \text{if } (i,j) \notin S. \end{cases}$$

This is an extension of the independence model that includes a *subtable change-point* parameter λ [53]. The sufficient statistics of this log-linear model are the row sums, columns sums, and S-subtable sum of a 2-way table u. For instance, if $r = 2$, and $c = 3$, and $S = \{(1,1), (2,2)\}$, then the sufficient statistics of this log-linear model are

$$
\begin{pmatrix}
1 & 1 & 1 & 0 & 0 & 0 \\
0 & 0 & 0 & 1 & 1 & 1 \\
1 & 0 & 0 & 1 & 0 & 0 \\
0 & 1 & 0 & 0 & 1 & 0 \\
0 & 0 & 1 & 0 & 0 & 1 \\
1 & 0 & 0 & 0 & 1 & 0
\end{pmatrix}
\begin{pmatrix}
u_{11} \\
u_{12} \\
u_{13} \\
u_{21} \\
u_{22} \\
u_{23}
\end{pmatrix}
$$

and the minimal Markov basis consists of a single move and its negative:

$$
\begin{pmatrix}
+1 & +1 & -2 \\
-1 & -1 & +2
\end{pmatrix}.
$$

1. Use the `markov` command in `4ti2` [57] to compute a Markov basis for the model when $r = c = 4$ and $S =$

   (a) $\{(1,1), (2,2), (3,3), (4,4)\}$,

   (b) $\{(1,1), (1,2), (2,1), (2,2)\}$,

   (c) $\{(1,1), (1,2), (2,1), (2,2), (3,3), (3,4), (4,3), (4,4)\}$,

   (d) $\{(1,1), (1,2), (1,3), (2,1), (2,2), (3,1)\}$.

2. Describe a Markov basis for this model when $S = S_1 \times S_2$ for $S_1 \subseteq [r]$, $S_2 \subseteq [c]$.

3. Describe a Markov basis for this model when $r = c$ and $S = \{(i,i) \mid i \in [r]\}$ is the set of diagonals.

Solution. 1. We consider the log-linear model for a 2-way contingency table with $r = c = 4$. The model is specified by the matrix A that consists of the restrictions from the independence model and an additional row depending on the set S. Our goal is to find a Markov basis of moves for each set S. These moves leave the column sums, rows sums, and the sum over S unchanged. Each move is an integer vector b of length $r \times c$ satisfying $Ab = 0$ but we can also view b as an $r \times c$ matrix, in which case the transpose of b is denoted b^T. Using the latter representation, let e_{ij} be the unit vector with a 1 at entry (i,j). Recall from Section 1.2 that the basic moves for the independence model involve only four elements and are of the form $e_{ik} + e_{jl} - e_{il} - e_{jk}$. This example of a basic move adds one to the cells (i,j) and (k,l) and subtracts one from the cells (i,l) and (j,k) in the contingency table.

  If b is a move, then so is $-b$. Note that the output of `4ti2` contains only one of the two vectors $\pm b$. Furthermore, note that `4ti2` always computes a minimal Markov basis.

  (a) In order to compute a Markov basis for $S = \{(1,1), (2,2), (3,3), (4,4)\}$ using the `markov` command in 4ti2, we set up the matrix

$$A = \begin{pmatrix}
1 & 1 & 1 & 1 & 0 & 0 & 0 & 0 & 0 & 0 & 0 & 0 & 0 & 0 & 0 & 0 \\
0 & 0 & 0 & 0 & 1 & 1 & 1 & 1 & 0 & 0 & 0 & 0 & 0 & 0 & 0 & 0 \\
0 & 0 & 0 & 0 & 0 & 0 & 0 & 0 & 1 & 1 & 1 & 1 & 0 & 0 & 0 & 0 \\
0 & 0 & 0 & 0 & 0 & 0 & 0 & 0 & 0 & 0 & 0 & 0 & 1 & 1 & 1 & 1 \\
1 & 0 & 0 & 0 & 1 & 0 & 0 & 0 & 1 & 0 & 0 & 0 & 1 & 0 & 0 & 0 \\
0 & 1 & 0 & 0 & 0 & 1 & 0 & 0 & 0 & 1 & 0 & 0 & 0 & 1 & 0 & 0 \\
0 & 0 & 1 & 0 & 0 & 0 & 1 & 0 & 0 & 0 & 1 & 0 & 0 & 0 & 1 & 0 \\
0 & 0 & 0 & 1 & 0 & 0 & 0 & 1 & 0 & 0 & 0 & 1 & 0 & 0 & 0 & 1 \\
1 & 0 & 0 & 0 & 0 & 1 & 0 & 0 & 0 & 0 & 1 & 0 & 0 & 0 & 0 & 1
\end{pmatrix}$$

that is passed as input to 4ti2. For each row of A, the 1's correspond to elements whose sum must be fixed. The first four rows correspond to row sums, the next four rows correspond to column sums, and the last row corresponds to the sum over the elements indexed by S.

The Markov basis returned by the `markov` command contains 66 moves. Using the formatting command `output`, we see that there are 6 elements of degree 2, 24 elements of degree 3, and 36 elements of degree 4. The elements of degree 2 in this Markov basis are basic moves $e_{ik} + e_{jl} - e_{il} - e_{jk}$ that do not change the sum over S. The other 60 elements can be written as the sum of two basic moves, where each one changes the sum over S but their sum does not. One can distinguish five types of Markov basis elements and we describe them for general $r = c$ in part 3.

(b) Now we consider $S = \{(1,1),(1,2),(2,1),(2,2)\}$, that is, the upper 2×2-rectangle of the table. Changing the last row of A according to the set S, we can again use the `markov` command to compute a Markov basis, which consists of 20 elements of degree 2. These are precisely the basic moves that do not change the sum over S.

(c) For $S = \{(1,1),(1,2),(2,1),(2,2),(3,3),(3,4),(4,3),(4,4)\}$, that is, the upper left and the lower right rectangle, we get the same Markov basis as for the upper left rectangle in part (b). The reason for this is that a basic move changes the sum over the upper left rectangle if and only if it changes the sum over the lower right rectangle.

(d) If $S = \{(1,1),(1,2),(1,3),(2,1),(2,2),(3,1)\}$ is the upper left triangle, then we find that the Markov basis computed by 4ti2 comprises 16 moves. As for the upper rectangle in part (b), these are all basic moves that do not change the sum of S.

We note that in [53, Thm. 1] it is proven that the set of allowable basic moves is a Markov basis for S if and only if S does not contain precisely the diagonal elements $(1,1)$ and $(2,2)$ in any 3×2 or 2×3 sub-matrix of $[r] \times [c]$. This condition is satisfied for the sets S in (b),(c),(d), while it is violated for the diagonal S in (a). The theorem is thus in agreement with our computational results.

2. Let $S_1 \subseteq [r], S_2 \subseteq [c]$ and consider $S = S_1 \times S_2$. Part (b) of question 1, above, is an example. Let $\mathcal{B}_\emptyset$ be the minimal Markov basis for the independence model, which contains the basic moves $e_{ik} + e_{jl} - e_{il} - e_{jk}$ for $i < j$ and $k < l$. We claim that if $S = S_1 \times S_2$, then the set

$$\mathcal{B}_S = \left\{ b \in \mathcal{B}_\emptyset \ : \ \sum_{(i,j) \in S} b_{ij} = 0 \right\} \tag{6.1.1}$$

is a Markov basis for the model with the additional restriction based on S. The set $\mathcal{B}_S$ comprises all basic moves that do not change the sum in S. Let $s_1 = \#S_1$ and $s_2 = \#S_2$. Then the number of moves in this basis is

$$\#\mathcal{B}_S = \binom{r}{2}\binom{c}{2} - s_1 s_2 (r - s_1)(c - s_2). \tag{6.1.2}$$

The claim holds according to [53, Thm. 1], but for the particular set S considered here we can give a simpler proof. Let u and v be two distinct non-negative integer matrices that have equal column sums, row sums, and sum over S. We need to show that one can reach v from u by applying a series of moves from $\mathcal{B}_S$ while keeping all matrix entries non-negative. As in the proof of Proposition 1.2.2 it suffices to show that we can find $b_1, \ldots, b_L \in \mathcal{B}_S$ such that $u + b_1 + \cdots + b_l \geq 0$ for all $1 \leq l \leq L$ and $\|u + b_1 + \cdots + b_L - v\|_1 < \|u - v\|_1$.

Suppose there exist two distinct indices $i, j \in [r] \setminus S_1$ with $u_{ik} - v_{ik} < 0$ and $u_{jk} - v_{jk} > 0$ for some $k \in [c]$. Then there is another column $l \in [c] \setminus \{k\}$ for which $u_{il} - v_{il} > 0$. We can thus reduce the one-norm by adding $e_{ik} + e_{jl} - e_{jk} - e_{il}$ to u. The same argument applies if $i, j \in S_1$ with $u_{ik} - v_{ik} < 0$ and $u_{jk} - v_{jk} > 0$ for some $k \in [c]$. Moreover, we can interchange the role of rows and columns.

In the remaining cases, there exist two distinct indices $i, j \in S_1$ and two distinct indices $k, l \in S_2$ with $u_{ik} - v_{ik} > 0$ and $u_{jl} - v_{jl} < 0$. In order for the above reasoning not to apply both $u_{il} - v_{il}$ and $u_{jk} - v_{jk}$ must be zero, and there must exist an index $m \in [c] \setminus S_2$ such that $u_{jm} - v_{jm} > 0$. Update u by adding the move $e_{jk} + e_{im} - e_{ik} - e_{jm}$. Then $u_{jk} - v_{jk} = 1$ and $u_{jl} - v_{jl} < 0$ is unchanged. We are done because we are back in the previously considered situation.

3. We are asked to describe a Markov basis when $S = \{(i, i) : i \in [r]\}$ is the diagonal in a square table. The case $r = c = 4$ was considered in part 1(a).

We represent each basic move by showing the non-zero elements participating in it. A composition of basic moves is represented similarly, where the number of non-zero elements might vary. For ease of presentation, we display moves in 4×4 matrices but they represent the same moves as in general $r \times r$ matrices. Elements appearing on the diagonal are underlined. We introduce five classes of moves. Recall that for each move in the Markov basis, we can use either it, or its negation. We always consider only one of these as a member of the Markov basis.

(i) The set $\mathcal{B}_1$ contains the $\binom{r}{2}\binom{r-2}{2}$ basic moves $e_{i_1j_1} + e_{i_2j_2} - e_{i_1j_2} - e_{i_2j_1}$ that do not contain the diagonal. An example is

$$\begin{pmatrix} +1 & -1 \\ -1 & +1 \end{pmatrix}.$$

(ii) The set $\mathcal{B}_2$ comprises $2\binom{r}{2}(r-2)$ moves created from two overlapping basic moves. In each of the two basic moves one of the elements is on the diagonal. The moves in $\mathcal{B}_2$ have the form $e_{i_1i_1} + e_{i_1i_2} - 2e_{i_1j_3} - e_{i_2i_1} - e_{i_2i_2} + 2e_{i_2j_3}$, possibly after transposition. An example is

$$\begin{pmatrix} \underline{+1} & +1 & -2 \\ -1 & \underline{-1} & +2 \end{pmatrix}.$$

(iii) The third set $\mathcal{B}_3$ is made up from pairs of non-overlapping basic moves in the same rows, each containing one diagonal element. Modulo transposition, the $2\binom{r}{2}\binom{r-2}{2}$ moves in $\mathcal{B}_3$ are of the form $e_{i_1i_1} + e_{i_1i_2} - e_{i_1j_3} - e_{i_1j_4} - e_{i_2i_1} - e_{i_2i_2} + e_{i_2j_3} + e_{i_2j_4}$, as illustrated by

$$\begin{pmatrix} \underline{+1} & +1 & -1 & -1 \\ -1 & \underline{-1} & +1 & +1 \end{pmatrix}.$$

(iv) The forth set $\mathcal{B}_4$ collects moves created from two overlapping basic moves, each containing two diagonal elements of which one is shared. They are of the form $e_{i_1i_1} - e_{i_1j_3} - e_{i_2i_2} + e_{i_2j_3} - e_{i_3i_1} + e_{i_3i_2}$, as illustrated by

$$\begin{pmatrix} \underline{+1} & & -1 \\ & \underline{-1} & +1 \\ -1 & +1 & \end{pmatrix}.$$

The indices are chosen distinct with the exception of i_3 and j_3, which may be equal. Based on this index structure, there are $\binom{r}{2}(r-2)^2$ moves in $\mathcal{B}_4$. However, some of these moves are redundant. In particular, when $i_3 = j_3$ we get a move that can be compensated by the two moves associated with the triplets (i_1, i_3, i_2) and (i_2, i_3, i_1). Excluding, for each triplet (i_1, i_2, i_3), one of the three moves in the basis, we obtain $\#\mathcal{B}_4 = \binom{r}{2}(r-2)^2 - \binom{r}{3}$.

(v) Finally, the set $\mathcal{B}_5$ is created from two overlapping basic moves, sharing their one diagonal element. These $\binom{r}{3}$ moves are of the form $e_{i_1 i_2} - e_{i_1 i_3} - e_{i_2 i_1} + e_{i_2 i_3} - e_{i_3 i_1} + e_{i_3 i_2}$. An example of an element of $\mathcal{B}_5$ is

$$\begin{pmatrix} & +1 & -1 \\ -1 & & +1 \\ -1 & +1 & \end{pmatrix}.$$

The union $\mathcal{B}_S = \bigcup_{i=1}^{5} \mathcal{B}_i$ of the above listed moves has cardinality

$$\binom{r}{2}\binom{r-2}{2} + 2\binom{r}{2}(r-2) + 2\binom{r}{2}\binom{r-2}{2} + \binom{r}{2}(r-2)^2 - \binom{r}{3} + \binom{r}{3}$$

$$= \frac{3}{2}\binom{r}{3}(5r-9),$$

which is equal to 9, 66, 240, and 630 for $r = 3, \ldots, 6$. Using 4ti2 we verified that $\mathcal{B}_S$ is a Markov basis for $r \leq 6$. This confirms the following result of [54].

Theorem. *Suppose* $S = \{(i,i) : i \in [r]\}$ *is the diagonal in an* $r \times r$*-contingency table. Then* $\mathcal{B}_S = \bigcup_{i=1}^{5} \mathcal{B}_i$ *is a Markov basis for the log-linear subtable change-point model associated with* S.

6.2 Quasi-symmetry and Cycles

The team consisted of Krzysztof Latuszynski and Carlos Trenado.

Problem. The *quasi-symmetry* model for two discrete random variables X, Y with the same number of states is the log-linear model with

$$\log p_{ij} = \alpha_i + \beta_j + \lambda_{ij}$$

where $\lambda_{ij} = \lambda_{ji}$.

1. What are the sufficient statistics of this model?

2. Compute the minimal Markov basis of the model for a few different values of the number of states of X, Y.

3. Give a combinatorial description of the minimal Markov basis of the quasi-symmetry model.

Solution. 1. Without loss of generality, assume that X and Y take values in the set $[r]$. For a sample $(X^{(1)}, Y^{(1)}), \ldots, (X^{(n)}, Y^{(n)})$, let $u = (u_{ij})_{i,j}$ be the data matrix, so $u_{ij} = \sum_{k=1}^{n} 1_{\{X^{(k)}=i, Y^{(k)}=j\}}$. To determine the sufficient statistics, we represent this model by a matrix A whose rows correspond to the parameters and whose columns correspond to the entries of the table u. The columns are labeled by pairs $(i,j) \in [r] \times [r]$ and the rows come in three blocks, corresponding to the α, β, and λ parameters. The first two blocks each have r rows (corresponding to rows and columns of u, respectively) and the λ block has $\binom{r+1}{2}$ rows which correspond to the 2 element multisets of $[r]$. The column corresponding to entry u_{ij} has three ones in it, corresponding to the three parameter types, and all other entries are zero. The positions of the ones are in the row corresponding to α_i, the row corresponding to β_j, and the row corresponding to $\lambda_{ij} = \lambda_{ji}$. For example if $r = 3$, the matrix A has the form:

$$
A = \left(
\begin{array}{ccccccccc}
1 & 1 & 1 & 0 & 0 & 0 & 0 & 0 & 0 \\
0 & 0 & 0 & 1 & 1 & 1 & 0 & 0 & 0 \\
0 & 0 & 0 & 0 & 0 & 0 & 1 & 1 & 1 \\
\hline
1 & 0 & 0 & 1 & 0 & 0 & 1 & 0 & 0 \\
0 & 1 & 0 & 0 & 1 & 0 & 0 & 1 & 0 \\
0 & 0 & 1 & 0 & 0 & 1 & 0 & 0 & 1 \\
\hline
1 & 0 & 0 & 0 & 0 & 0 & 0 & 0 & 0 \\
0 & 0 & 0 & 0 & 1 & 0 & 0 & 0 & 0 \\
0 & 0 & 0 & 0 & 0 & 0 & 0 & 0 & 1 \\
0 & 1 & 0 & 1 & 0 & 0 & 0 & 0 & 0 \\
0 & 0 & 1 & 0 & 0 & 0 & 1 & 0 & 0 \\
0 & 0 & 0 & 0 & 0 & 1 & 0 & 1 & 0 \\
\end{array}
\right) .
$$

Now to determine the sufficient statistics of the quasi-symmetry model from the matrix A, we read across the rows of A. The α and β blocks correspond to row and column sums of u, as in a typical independence model. For the λ block, this gets broken into two classes: diagonal and off-diagonal. If we are looking at λ_{ii}, u_{ii} is the only table entry affected by this parameter, in which case the corresponding sufficient statistic is u_{ii}. For off-diagonal λ_{ij}, u_{ij} and u_{ji} are affected, and the vector of sufficient statistics is

$$
\begin{aligned}
u_{i+} \quad &\text{for} \quad i = 1, \ldots, r, \\
u_{+j} \quad &\text{for} \quad j = 1, \ldots, r, \\
u_{ii} \quad &\text{for} \quad i = 1, \ldots, r, \\
u_{ij} + u_{ji} \quad &\text{for} \quad i, j = 1, \ldots, r \text{ and } i < j.
\end{aligned}
$$

2. To compute the Markov basis for the model for $r \times r$ tables we input the matrix that computes sufficient statistics of the model to 4ti2 [57]. That matrix, which is an $2r + \binom{r+1}{2}$ by r^2 matrix, was described in the solution to part 1.

For $r = 3$, the Markov basis consists of the 2 moves

$$
\pm \left(
\begin{array}{ccc}
0 & 1 & -1 \\
-1 & 0 & 1 \\
1 & -1 & 0
\end{array}
\right) .
$$

For $r = 4$, the Markov basis consists of $2 \cdot 7$ moves, given by the following matrices:

$$\pm \begin{pmatrix} 0 & 0 & 0 & 0 \\ 0 & 0 & -1 & +1 \\ 0 & +1 & 0 & -1 \\ 0 & -1 & +1 & 0 \end{pmatrix}, \pm \begin{pmatrix} 0 & 0 & -1 & +1 \\ 0 & 0 & 0 & 0 \\ +1 & 0 & 0 & -1 \\ -1 & 0 & +1 & 0 \end{pmatrix}, \pm \begin{pmatrix} 0 & -1 & 0 & +1 \\ +1 & 0 & 0 & -1 \\ 0 & 0 & 0 & 0 \\ -1 & +1 & 0 & 0 \end{pmatrix},$$

$$\pm \begin{pmatrix} 0 & 0 & -1 & +1 \\ 0 & 0 & +1 & -1 \\ +1 & -1 & 0 & 0 \\ -1 & +1 & 0 & 0 \end{pmatrix}, \pm \begin{pmatrix} 0 & -1 & 0 & +1 \\ +1 & 0 & -1 & 0 \\ 0 & +1 & 0 & -1 \\ -1 & 0 & +1 & 0 \end{pmatrix},$$

$$\pm \begin{pmatrix} 0 & -1 & +1 & 0 \\ +1 & 0 & -1 & 0 \\ -1 & +1 & 0 & 0 \\ 0 & 0 & 0 & 0 \end{pmatrix}, \pm \begin{pmatrix} 0 & +1 & -1 & 0 \\ -1 & 0 & 0 & +1 \\ +1 & 0 & 0 & -1 \\ 0 & -1 & +1 & 0 \end{pmatrix}.$$

We leave to the reader the enjoyable task of designing the A matrix and computing the $2 \cdot 37$ moves of the Markov basis for $r = 5$ using 4ti2.

3. Markov bases for the quasi-symmetry model were considered e.g. in [76], where the 4×4 example has been computed. However, the problem was not solved for arbitrary r. Here we solve the problem of determining the minimal Markov basis of this model by providing a proof of a conjecture by Filip Cools.

To each cyclic permutation σ of $[r]$ of length between 3 to r, we associate an $r \times r$ table $M_\sigma = (m_{ij})$, where

$$m_{ij} = \begin{cases} 1 & \text{if} \quad \sigma(i) = j, \\ -1 & \text{if} \quad \sigma^{-1}(i) = j, \\ 0 & \text{otherwise.} \end{cases}$$

Note that $M_{\sigma^{-1}} = -M_\sigma$.

Theorem. *The minimal Markov basis $\mathcal{B}$ of the $r \times r$ quasi-symmetry model is the union of all the moves M_σ as σ ranges over all cyclic permutations with cycle length between 3 and r. In particular, the Markov basis consists of*

$$\sum_{k=3}^{r} \binom{r}{k} (k-1)! \tag{6.2.1}$$

moves.

Proof. First of all note that $M_\sigma \in \ker_{\mathbb{Z}} A$. It is easy to verify that the number of such cycles, and thus the number of such matrices, is given by (6.2.1).

We must show that if we have two non-negative $r \times r$ integral tables u, v with the same row and column sums, same diagonals, and same symmetric pair sum, they can be connected to each other using the set of cyclic moves $\mathcal{B}$. Using the same strategy as in the proof of Proposition 1.2.2, it suffices to show that there is a $b \in \mathcal{B}$ such that $u + b \geq 0$ and $\|u + b - v\|_1 < \|u - v\|_1$.

Suppose that $u \neq v$. Then there exists some index pair (i_1, i_2) such that $u_{i_1 i_2} - v_{i_1 i_2} > 0$. By virtue of the fact that u and v have the same symmetric pair sums we have that $u_{i_2 i_1} - v_{i_2 i_1} < 0$. Since u and v have the same column sums, there is an index i_3 such that $u_{i_2 i_3} - v_{i_2 i_3} > 0$. By virtue of the fact that u and v have the same symmetric pair sums we have that $u_{i_3 i_2} - v_{i_3 i_2} < 0$. Since u and v have the same column sums, there is an index i_4 such that $u_{i_3 i_4} - v_{i_3 i_4} > 0$. Continuing in this way for at most r steps, we produce a sequence of indices $i_1 i_2 \cdots i_{k+1}$ with $i_{k+1} = i_1$ such that $u_{i_l i_{l+1}} - v_{i_l i_{l+1}} > 0$ and $u_{i_{l+1} i_l} - v_{i_{l+1} i_l} < 0$ for all $l = 1, 2, \ldots, k$. Letting σ be the corresponding cyclic permutation $\sigma = (i_1 i_2 \cdots i_k)$ and $b = M_{\sigma^{-1}}$ gives the desired move.

Applying the same connecting argument to the pair of tables M_σ^+ and M_σ^-, the positive and negative parts of M_σ, shows that none of these moves can be omitted from the Markov basis, and hence, this is a minimal Markov basis. $\square$

Note that the argument in the preceding proof shows that the set of moves is more than just a Markov basis for the model, they form a Graver basis as well.

6.3 A Colored Gaussian Graphical Model

The team consisted of Florian Block, Sofia Massa and Martina Kubitzke.

Problem. Let $\Theta \subset PD_m$ be the set of positive definite $m \times m$-matrices that can be written as ACA, where A is a diagonal matrix with diagonal entries $\alpha_1, \ldots, \alpha_m > 0$ and

$$
C = \begin{pmatrix}
1 & \gamma & & & \\
\gamma & 1 & \gamma & & \\
& \ddots & \ddots & \ddots & \\
& & \gamma & 1 & \gamma \\
& & & \gamma & 1
\end{pmatrix}
$$

is a tridiagonal positive definite matrix. Consider the model of all multivariate normal distributions $\mathcal{N}(\mu, \Sigma)$ with $\mu \in \mathbb{R}^m$ and *concentration matrix* $\Sigma^{-1} \in \Theta$. This model is an instance of a *colored Gaussian graphical model* [60].

1. Suppose we observe a positive definite sample covariance matrix $S = (s_{ij})$. Show that the likelihood function involves only s_{ij} with $|i - j| \leq 1$, and that for solution of the likelihood equations we may without loss of generality replace S by the sample correlation matrix $R = (r_{ij})$ which has the entries

$$
r_{ij} = \begin{cases}
1 & \text{if } i = j, \\
\dfrac{s_{ij}}{\sqrt{s_{ii} s_{jj}}} & \text{if } i \neq j.
\end{cases}
$$

2. Compute all solutions to the likelihood equations if $m = 3$ and

$$S = \begin{pmatrix} 7 & 2 & -1 \\ 2 & 5 & 3 \\ -1 & 3 & 11 \end{pmatrix}.$$

3. Study the likelihood equations for $m = 3$, treating the correlations r_{12} and r_{23} as parameters. Can there be more than one feasible solution?

4. Compute ML degrees for some $m \geq 4$.

Solution. 1. In (2.1.5) in Section 2.1, we saw that the log-likelihood function for a Gaussian model is

$$\ell_n(\mu, \Sigma) = -\frac{n}{2} \log \det \Sigma - \frac{n}{2} \mathrm{tr}(S\Sigma^{-1}) - \frac{n}{2}(\bar{X} - \mu)^T \Sigma^{-1}(\bar{X} - \mu),$$

where n is the sample size. We also saw that the ML estimator of μ is the sample mean $\bar{X}$ and that for estimation of Σ we can maximize the function

$$\ell(\Sigma) = -\log \det \Sigma - \mathrm{tr}(S\Sigma^{-1}). \tag{6.3.1}$$

If $\Sigma^{-1} = ACA$, then $\ell(\Sigma)$ is equal to

$$2 \log \det A + \log \det C - \mathrm{tr}(ASAC). \tag{6.3.2}$$

For $i, j \in [m]$, let E_{ij} be the zero-one matrix that has 1s at exactly the entries (i, j) and (j, i). If $i = j$, then E_{ii} has only the entry (i, i) equal to 1. Then

$$\mathrm{tr}(ASAE_{ii}) = \alpha_i^2 s_{ii} \quad \text{and} \quad \mathrm{tr}(ASAE_{ij}) = 2\alpha_i \alpha_j s_{ij},$$

and since

$$C = \gamma \sum_{i=1}^{m-1} E_{i,i+1} + \sum_{i=1}^{m} E_{ii}$$

it holds that

$$\mathrm{tr}(ASAC) = \gamma \sum_{i=1}^{m-1} 2\alpha_i \alpha_{i+1} s_{i,i+1} + \sum_{i=1}^{m} \alpha_i^2 s_{ii}$$

involves only the sample covariances s_{ij} with $|i - j| \leq 1$.

Let D_S be the diagonal matrix with $\sqrt{s_{ii}}$ as i-th diagonal entry. Then $S = D_S R D_S$. Using the invertible linear transformation $A \mapsto \bar{A} = A D_S$, (6.3.2) can be rewritten as

$$2 \log \det \bar{A} + \log \det C - \mathrm{tr}(\bar{A} R \bar{A} C) - 2 \log \det D_S. \tag{6.3.3}$$

Since $2 \log \det D_S$ depends only on the data S, we can find the critical points $(\alpha_1, \ldots, \alpha_m, \gamma)$ of (6.3.2) by computing the critical points $(\bar{\alpha}_1, \ldots, \bar{\alpha}_m, \gamma)$ of

$$2 \log \det \bar{A} + \log \det C - \operatorname{tr}(\bar{A} R \bar{A} C) \tag{6.3.4}$$

and transform the solutions $\bar{A}$ to $A = \bar{A} D_S^{-1}$, that is, we divide each $\bar{\alpha}_i$ by $\sqrt{s_{ii}}$.

2. The likelihood equations are obtained by setting the partial derivatives of (6.3.2) with respect to γ and $\alpha_1, \ldots, \alpha_m$ equal to zero. These are

$$\frac{\partial \ell(\Sigma)}{\partial \gamma} = \frac{1}{\det C} \cdot \frac{\partial \det C}{\partial \gamma} - \sum_{i=1}^{m-1} 2\alpha_i \alpha_{i+1} s_{i,i+1},$$

$$\frac{\partial \ell(\Sigma)}{\partial \alpha_i} = \frac{2}{\alpha_i} - 2\alpha_i s_{ii} - 2\gamma(\alpha_{i-1} s_{i-1,i} + \alpha_{i+1} s_{i,i+1}), \quad i = 1, \ldots, m,$$

where $s_{m,m+1}$ and $s_{0,1}$ are understood to be zero. Clearing denominators by multiplying by $\det C$ and α_i, respectively, and dividing each equation by 2, we obtain the polynomial equation system

$$\frac{1}{2} \frac{\partial \det C}{\partial \gamma} - \det C \sum_{i=1}^{m-1} \alpha_i \alpha_{i+1} s_{i,i+1} = 0, \tag{6.3.5a}$$

$$1 - \alpha_i^2 s_{ii} - \gamma \alpha_i (\alpha_{i-1} s_{i-1,i} + \alpha_{i+1} s_{i,i+1}) = 0, \quad i = 1, \ldots, m. \tag{6.3.5b}$$

We set up the equations for the given sample covariance matrix S in Singular:

```
LIB "solve.lib";
ring R = 0, (g,a(1..3)), lp;
matrix S[3][3] = 7,2,-1, 2,5,3, -1,3,11;
matrix C[3][3] = 1,g,0, g,1,g, 0,g,1;
ideal I = 1/2*diff(det(C),g)-det(C)*a(2)*(a(1)*S[1,2]+a(3)*S[2,3]),
          1-a(1)^2*S[1,1]-g*a(1)*a(2)*S[1,2],
          1-a(2)^2*S[2,2]-g*a(2)*(a(1)*S[1,2]+a(3)*S[2,3]),
          1-a(3)^2*S[3,3]-g*a(2)*a(3)*S[2,3];
```

When clearing denominators it is possible to introduce new solutions with α_i or $\det C$ equal to zero. However, this is not the case here. Clearly, all solutions to (6.3.5a) and (6.3.5b) have $\alpha_i \neq 0$, and since the command

```
groebner(I+ideal(det(C)));
```

returns 1, no solution satisfies $\det C = 0$. To compute the solutions to the likelihood equations we issue the command

```
solve(I);
```

There are eight solutions, all of them real, but only one is feasible having all $\alpha_i > 0$ and $\det C = 1 - 2\gamma^2 > 0$. The solutions come in two classes represented by

[4]: [1]: 0.035667176   [8]: [1]: -0.3482478
     [2]: 0.37568976   [2]: 0.40439204
     [3]: 0.44778361   [3]: 0.51385299
     [4]: -0.3036971   [4]: 0.32689921

where the components correspond to $(\gamma, \alpha_1, \alpha_2, \alpha_3)$. The other solutions are obtained by certain sign changes.

3. Continuing with the case $m = 3$, we use the observation that the solutions of the likelihood equations can be found by computing the critical points of (6.3.4), which only depends on r_{12} and r_{23}. Treating these correlations as parameters in Singular we define the ring and data as

```
ring R = (0,r12,r23), (g,a(1..3)), lp;
matrix S[3][3] = 1,r12,0, r12,1,r23, 0,r23,1;
```

The likelihood equations can then be set up using the code above, and we can compute a reduced lexicographic Gröbner basis by setting the option option(redSB) and typing groebner(I). The Gröbner basis corresponds to four equations:

$$(2 - r_{23}^2)(2 - r_{12}^2 - r_{23}^2) \cdot \alpha_3^4 - 2(2 - r_{23}^2)(2 - r_{12}^2) \cdot \alpha_3^2 + 2(2 - r_{12}^2) = 0,$$
$$(2 - r_{12}^2) \cdot \alpha_2^2 - (4 - r_{12}^2 - r_{23}^2) \cdot \alpha_3^2 + (2 - r_{12}^2) = 0,$$
$$-r_{12}r_{23}(2 - r_{12}^2) \cdot \alpha_1 + (2 - r_{23}^2)(2 - r_{12}^2 - r_{23}^2) \cdot \alpha_3^3 - (2 - r_{23}^2)(2 - r_{12}^2) \cdot \alpha_3 = 0,$$
$$2r_{23}h_1 \cdot \gamma + (2 - r_{23}^2)(2 - r_{12}^2 - r_{23}^2)h_2 \cdot \alpha_2\alpha_3^3 - 2(2 - r_{23}^2)h_3 \cdot \alpha_2\alpha_3 = 0,$$

where h_1, h_2, h_3 are polynomials in r_{12}^2 and r_{23}^2. In particular, the ML estimates can be computed by solving quadratic equations.

If $r_{12}^2, r_{23}^2 \leq 1$, as is necessarily the case if the sample covariance matrix S is positive definite, then a standard calculation shows that the first equation has four real solutions for α_3. All of them are such that the second equation has two real solutions for α_2. Hence, for a generic positive definite sample covariance matrix, the ML degree is eight and all eight solutions to the likelihood equations are real.

If the sample covariance matrix S is positive definite, then the function $\ell(\Sigma)$ in (6.3.1) is bounded over the positive definite cone and tends to minus infinity if Σ or Σ^{-1} approach a positive semi-definite matrix. It follows that the likelihood equations based on r_{12} and r_{23} have at least one feasible solution $(\alpha_{10}, \alpha_{20}, \alpha_{30}, \gamma_0)$. Similarly, the likelihood equations based on $-r_{12}$ and r_{23} have a feasible solution $(\alpha_{11}, \alpha_{21}, \alpha_{31}, \gamma_1)$. The Gröbner basis displayed above reveals that the eight real solutions to the likelihood equations based on r_{12} and r_{23} are

$$\begin{pmatrix} \alpha_{10} \\ \alpha_{20} \\ \alpha_{30} \\ \gamma_0 \end{pmatrix}, \begin{pmatrix} \alpha_{10} \\ -\alpha_{20} \\ \alpha_{30} \\ -\gamma_0 \end{pmatrix}, \begin{pmatrix} -\alpha_{10} \\ \alpha_{20} \\ -\alpha_{30} \\ -\gamma_0 \end{pmatrix}, \begin{pmatrix} -\alpha_{10} \\ -\alpha_{20} \\ -\alpha_{30} \\ \gamma_0 \end{pmatrix}$$

and

$$\begin{pmatrix} -\alpha_{11} \\ \alpha_{21} \\ \alpha_{31} \\ \gamma_1 \end{pmatrix}, \begin{pmatrix} -\alpha_{11} \\ -\alpha_{21} \\ \alpha_{31} \\ -\gamma_1 \end{pmatrix}, \begin{pmatrix} \alpha_{11} \\ \alpha_{21} \\ -\alpha_{31} \\ -\gamma_1 \end{pmatrix}, \begin{pmatrix} \alpha_{11} \\ -\alpha_{21} \\ -\alpha_{31} \\ \gamma_1 \end{pmatrix}.$$

Note that since $\alpha_{i0}, \alpha_{i1} > 0$ all eight sign combinations for the α_i occur. We conclude that, for a generic positive definite sample covariance matrix, the likelihood equations have exactly one feasible solution.

4. Using the code below we compute the ML degree for $m \geq 4$ (and confirm that the clearing of denominators did not introduce additional solutions). We generate data at random and for $m \geq 6$ we used a computation in different finite characteristics (e.g., set int c = 99991).

```
LIB "linalg.lib";
int m = 4; int c = 0;
intmat X = random(31,m,m);  intmat S = X*transpose(X);
ring R = c,(g,a(1..m)),lp;
matrix A[m][m];
for(int i=1;i<=m;i++){  A[i,i]=a(i); }
matrix C[m][m];
C[1,1] = 1;
for(i=2;i<=m;i++){  C[i,i] = 1; C[i-1,i]=g; C[i,i-1]=g; }
ideal I = diff(det(C),g)-diff(trace(A*C*A*S),g)*det(C);
for(i=1;i<=m;i++){
  I = I+ideal(1-1/2*diff(trace(A*C*A*S),a(i))*a(i));
}
ideal G = groebner(I);
dim(G);   vdim(G);
groebner(G+ideal(det(C)));
```

We obtain the results

m	3	4	5	6	7	8
ML deg	8	64	152	480	1072	2816

but, unfortunately, no obvious guess for a formula emerges.

6.4 Instrumental Variables and Tangent Cones

The team consisted of Shaowei Lin, Thomas Kahle and Oliver Wienand.

Problem. Let $\varepsilon = (\varepsilon_1, \ldots, \varepsilon_4) \sim \mathcal{N}(0, \Omega)$ be a multivariate normal random vector with positive definite covariance matrix

$$\Omega = \begin{pmatrix} \omega_{11} & 0 & 0 & 0 \\ 0 & \omega_{22} & 0 & 0 \\ 0 & 0 & \omega_{33} & \omega_{34} \\ 0 & 0 & \omega_{34} & \omega_{44} \end{pmatrix}. \tag{6.4.1}$$

Define new random variables $X_1, \ldots, X_4$ as linear combinations:

$$X_1 = \varepsilon_1, \qquad\qquad X_3 = \lambda_{31} X_1 + \lambda_{32} X_2 + \varepsilon_3,$$
$$X_2 = \varepsilon_2, \qquad\qquad X_4 = \lambda_{43} X_3 + \varepsilon_4.$$

Then $X_1, \ldots, X_4$ have zero means and are jointly multivariate normal. Let Θ be the set of covariance matrices of $X = (X_1, \ldots, X_4)$ obtained by choosing any real numbers for the coefficients λ_{ij} and a positive definite matrix Ω as in (6.4.1).

1. Write the covariance matrix $\Sigma = (\sigma_{ij})$ of X as a function of the entries of Ω and the coefficients λ_{ij}.

2. Describe the ideal I of polynomials in $\mathbb{R}[s_{ij} \mid 1 \leq i \leq j \leq 4]$ that evaluate to zero at all matrices in Θ. What is the singular locus of the variety $V(I)$. What choices of coefficients λ_{ij} lead to singularities?

3. Let $\Sigma \in \Theta$ be a singularity of $V(I)$. What is the tangent cone of Θ at Σ?

4. Find all possible asymptotic distributions of the likelihood ratio statistic λ_n for testing

$$H_0 : \Sigma \in \Theta \quad \text{versus} \quad H_1 : \Sigma \notin \Theta$$

   when the true distribution $\mathcal{N}(0, \Sigma_0)$ has $\Sigma_0 \in \Theta$.

Solution. 1. Define the matrix

$$\Lambda = \begin{pmatrix} 1 & 0 & 0 & 0 \\ 0 & 1 & 0 & 0 \\ -\lambda_{31} & -\lambda_{32} & 1 & 0 \\ 0 & 0 & -\lambda_{43} & 1 \end{pmatrix}.$$

Then $X = \Lambda^{-1} \varepsilon$ has covariance matrix $\Sigma = \Lambda^{-1} \Omega \Lambda^{-T}$, which is equal to

$$\begin{pmatrix} \omega_{11} & 0 & \lambda_{31}\omega_{11} & \lambda_{31}\lambda_{43}\omega_{11} \\ & \omega_{22} & \lambda_{32}\omega_{22} & \lambda_{32}\lambda_{43}\omega_{22} \\ & & \omega_{33} + \lambda_{31}^2\omega_{11} + \lambda_{32}^2\omega_{22} & \omega_{34} + \lambda_{43}\sigma_{33} \\ & & & \omega_{44} + 2\lambda_{43}\omega_{34} + \lambda_{43}^2\sigma_{33} \end{pmatrix} \tag{6.4.2}$$

where the symmetric entries below the diagonal are omitted and the shorthand $\sigma_{33} = \omega_{33} + \lambda_{31}^2\omega_{11} + \lambda_{32}^2\omega_{22}$ is used in the fourth column.

Remark. We see that despite the possible dependence of the error terms ε_3 and ε_4 the coefficient λ_{43} can be obtained from Σ using the *instrumental variables* formulas

$$\lambda_{43} = \frac{\sigma_{j4}}{\sigma_{j3}}, \quad j = 1, 2.$$

These formulas require σ_{13} or σ_{23} (or equivalently, λ_{31} or λ_{32}) to be non-zero.

2. It is easy to see that if $\Sigma = (\sigma_{ij})$ is the covariance matrix of X, then it holds that $\sigma_{12} = 0$ and $\sigma_{13}\sigma_{24} - \sigma_{14}\sigma_{23} = 0$. In fact, all relations between entries of the matrix are given by the ideal

$$I = \langle \sigma_{13}\sigma_{24} - \sigma_{14}\sigma_{23}, \sigma_{12} \rangle. \tag{6.4.3}$$

The next piece of Singular code verifies this claim:

```
LIB "linalg.lib";
ring R = 0,(w1,w2,w3,w34,w4, l31,l32,l43,
              s11,s12,s13,s14, s22,s23,s24, s33,s34, s44),dp;
matrix L[4][4] = 1,0,0,0,
                 0,1,0,0,
                 -l31,-l32,1,0,
                 0,0,-l43,1;
matrix W[4][4] = w1,0,0,0,
                 0,w2,0,0,
                 0,0,w3,w34,
                 0,0,w34,w4;
matrix Sigma[4][4] = inverse(L)*W*transpose(inverse(L));
matrix S[4][4] = s11,s12,s13,s14,
                 s12,s22,s23,s24,
                 s13,s23,s33,s34,
                 s14,s24,s34,s44;
ideal E = 0;
for(int i=1; i<=4; i++){
  for(int j=1; j<=4; j++){
     E = E+ideal(S[i,j]-Sigma[i,j]);
  }
}
ideal I = eliminate(E,w1*w2*w3*w34*w4*l31*l32*l43);
I;
```

In order to find the singularities we study the rank of the Jacobian J of the implicit equations, which is

$$J = \begin{pmatrix} 1 & 0 & 0 & 0 & 0 \\ 0 & \sigma_{24} & -\sigma_{23} & -\sigma_{14} & \sigma_{13} \end{pmatrix};$$

recall Definition 2.3.15. The rank of J drops if and only if $\sigma_{13} = \sigma_{14} = \sigma_{23} = \sigma_{24} = 0$, which is the case if and only if $\lambda_{31} = \lambda_{32} = 0$ (recall that $\omega_{11}, \omega_{22} > 0$). For computation of the singularities using `Singular`, load the library LIB `"sing.lib"` and type `slocus(I)` or for a nicer output `groebner(slocus(I))`.

3. Let g be the parametrization map, which maps a vector of parameters

$$(\omega, \lambda) = (\omega_{11}, \omega_{22}, \omega_{33}, \omega_{34}, \omega_{44}, \lambda_{31}, \lambda_{32}, \lambda_{43}) \tag{6.4.4}$$

to the covariance matrix displayed in (6.4.2). According to part 2, a matrix $\bar{\Sigma} = g(\bar{\omega}, \bar{\lambda})$ is a singularity of Θ if and only if $\bar{\lambda}_{31} = \bar{\lambda}_{32} = 0$, in which case it is of the form

$$\bar{\Sigma} = \begin{pmatrix} \bar{\omega}_{11} & 0 & 0 & 0 \\ & \bar{\omega}_{22} & 0 & 0 \\ & & \bar{\omega}_{33} & \bar{\omega}_{34} + \bar{\lambda}_{43}\bar{\omega}_{33} \\ & & & \bar{\omega}_{44} + 2\bar{\lambda}_{43}\bar{\omega}_{34} + \bar{\lambda}_{43}^2\bar{\omega}_{33} \end{pmatrix}. \tag{6.4.5}$$

By Lemma 2.3.17, the tangent cone $TC_{\bar{\Sigma}}(\Theta)$ is a subset of the variety $V(I)$, which is itself a cone. We claim that $TC_{\bar{\Sigma}}(\Theta) = V(I)$. To confirm this we study the Jacobian $H(\omega, \lambda)$ of the parametrization g at a singular point. Continuing the `Singular` session from part 2, we compute

```
matrix H[10][8];
int i,j,c; int r = 0;
for(i=1;i<=4;i++){
  for(j=i;j<=4;j++){
    r = r+1;
    for(c=1;c<=8;c++){
      H[r,c] = diff(Sigma[i,j],var(c));
    }
  }
}
```

When stating the result we vectorize the covariance matrices $\Sigma = (\sigma_{ij})$ in the image of the parametrization as

$$(\sigma_{12}, \sigma_{11}, \sigma_{22}, \ \sigma_{33}, \sigma_{34}, \sigma_{44}, \ \sigma_{13}, \sigma_{14}, \sigma_{23}, \sigma_{24}) \tag{6.4.6}$$

and list partial derivatives in the order given by (6.4.4). Then the Jacobian $H(\omega, \lambda)$

when $\lambda_{31} = \lambda_{32} = 0$ is

$$H(\bar{\omega}, \bar{\lambda}) = \begin{pmatrix} 0 & 0 & 0 & 0 & 0 & 0 & 0 & 0 \\ 1 & 0 & 0 & 0 & 0 & 0 & 0 & 0 \\ 0 & 1 & 0 & 0 & 0 & 0 & 0 & 0 \\ 0 & 0 & 1 & 0 & 0 & 0 & 0 & 0 \\ 0 & 0 & \bar{\lambda}_{43} & 1 & 0 & 0 & 0 & \bar{\omega}_{33} \\ 0 & 0 & \bar{\lambda}_{43}^2 & 2\bar{\lambda}_{43} & 1 & 0 & 0 & 2\bar{\omega}_{34} + 2\bar{\lambda}_{43}\bar{\omega}_{33} \\ 0 & 0 & 0 & 0 & 0 & \bar{\omega}_{11} & 0 & 0 \\ 0 & 0 & 0 & 0 & 0 & \bar{\lambda}_{43}\bar{\omega}_{11} & 0 & 0 \\ 0 & 0 & 0 & 0 & 0 & 0 & \bar{\omega}_{22} & 0 \\ 0 & 0 & 0 & 0 & 0 & 0 & \bar{\lambda}_{43}\bar{\omega}_{22} & 0 \end{pmatrix}.$$

According to Proposition 2.3.14, $TC_{\bar{\Sigma}}(\Theta)$ contains the column span of $H(\bar{\omega}, \bar{\lambda})$, which is the linear space

$$\{0\} \times \mathbb{R}^5 \times \mathcal{L}(\bar{\lambda}_{43}),$$

where

$$\mathcal{L}(\alpha) = \{(x_{13}, x_{14}, x_{23}, x_{24}) \in \mathbb{R}^4 \ : \ x_{14} = \alpha x_{13}, \ x_{24} = \alpha x_{23}\}.$$

However, the parametrization map g is not injective at a point (ω, λ) with $\lambda_{31} = \lambda_{32} = 0$. Indeed the preimage of $\bar{\Sigma} = (\bar{\sigma}_{ij})$ in (6.4.5) is one-dimensional and equals

$$g^{-1}(\bar{\Sigma}) = \{(\omega, \lambda) \in \mathbb{R}^8 \ : \ \omega_{11} = \bar{\sigma}_{11}, \ \omega_{22} = \bar{\sigma}_{22}, \ \omega_{33} = \bar{\sigma}_{33}, \ \lambda_{31} = \lambda_{32} = 0,$$
$$\omega_{34} = \bar{\sigma}_{34} - \lambda_{43}\bar{\sigma}_{33}, \ \omega_{44} = \bar{\sigma}_{44} - 2\lambda_{43}\bar{\sigma}_{34} + \lambda_{43}^2\bar{\sigma}_{33}\}.$$

Note that, since $\bar{\Sigma}$ is positive definite,

$$\omega_{44} = \bar{\sigma}_{44} - 2\lambda_{43}\bar{\sigma}_{34} + \lambda_{43}^2\bar{\sigma}_{33} > 0$$

and

$$\omega_{33}\omega_{44} - \omega_{34}^2 = \bar{\sigma}_{33}(\bar{\sigma}_{44} - 2\lambda_{43}\bar{\sigma}_{34} + \lambda_{43}^2\bar{\sigma}_{33}) - (\bar{\sigma}_{34} - \lambda_{43}\bar{\sigma}_{33})^2$$
$$= \bar{\sigma}_{33}\bar{\sigma}_{44} - \bar{\sigma}_{34}^2 > 0.$$

The coefficient λ_{43} in the parametrization of the preimage $g^{-1}(\bar{\Sigma})$ is arbitrary. Therefore, and since tangent cones are closed, $TC_{\bar{\Sigma}}(\Theta)$ contains the closure of

$$\{0\} \times \mathbb{R}^5 \times \bigcup_{\alpha \in \mathbb{R}} \mathcal{L}(\alpha).$$

This closure is equal to $V(I)$. Viewed as a variety in the space of symmetric matrices, $V(I)$ comprises all symmetric 4×4-matrices $\tau = (\tau_{ij})$ such that $\tau_{12} = 0$ and the submatrix $\tau_{12 \times 34}$ has rank at most 1; recall (6.4.3).

4. Let us first discuss the case of the true parameter Σ_0 being a non-singular point of Θ, that is, $\Sigma_0 = g(\omega, \lambda)$ with $\lambda_{31} \neq 0$ or $\lambda_{32} \neq 0$. Continuing the **Singular** session from part 3, we compute the locus where the rank of H is smaller than 8:

```
sat( minor(H,8), w1*w2*w3*w4);
```

where the saturation takes into account that $w_i > 0$. The output shows that the Jacobian H has rank smaller than 8 exactly when $\lambda_{31} = \lambda_{32} = 0$. By Lemma 2.3.16, it thus holds at all non-singular points Σ_0 that

$$\lambda_n \xrightarrow{D} \chi^2_{\operatorname{codim}(\Theta)} = \chi^2_2.$$

Next suppose that $\Sigma_0 = (\sigma_{ij})$ is a singularity of Θ. Chernoff's Theorem (Theorem 2.3.12) states that the asymptotic distribution of λ_n is given by

$$\min_{\tau \in TC_{\Sigma_0}(\Theta)} \|Z - I(\Sigma_0)^{1/2}\tau\|^2,$$

where $I(\Sigma_0) = I(\Sigma_0)^{T/2}I(\Sigma_0)^{1/2}$ is the 10×10-Fisher information matrix obtained from considering only covariances σ_{ij} with $1 \leq i \leq j \leq 4$, the random vector $Z \in \mathbb{R}^{10}$ is distributed according to $\mathcal{N}(0, Id)$, and $TC_{\Sigma_0}(\Theta)$ is the tangent cone of Θ at Σ_0 viewed as a subset of $\mathbb{R}^{10}$.

As shown in Exercise 6.5, the Fisher information matrix satisfies

$$(I(\Sigma_0)^{-1})_{ij,kl} = \sigma_{ik}\sigma_{jl} + \sigma_{il}\sigma_{jk}.$$

By (6.4.5), a singularity Σ_0 is of the form

$$\Sigma_0 = \begin{pmatrix} \sigma_{11} & 0 & 0 & 0 \\ & \sigma_{22} & 0 & 0 \\ & & \sigma_{33} & \sigma_{34} \\ & & & \sigma_{44} \end{pmatrix}.$$

The many zeros lead to a block-diagonal structure for the Fisher-information. When we order the rows and columns according to the variable ordering in (6.4.6), then

$$I(\Sigma_0)^{-1} = \begin{pmatrix} B_1 & & & \\ & B_2 & & \\ & & B_3 & \\ & & & B_4 \end{pmatrix}$$

where

$$B_1 = \begin{pmatrix} \sigma_{11}\sigma_{22} & & \\ & 2\sigma_{11}^2 & \\ & & 2\sigma_{22}^2 \end{pmatrix}, \quad B_2 = \begin{pmatrix} 2\sigma_{33}^2 & 2\sigma_{33}\sigma_{34} & 2\sigma_{34}^2 \\ 2\sigma_{33}\sigma_{34} & \sigma_{34}^2 + \sigma_{33}\sigma_{44} & 2\sigma_{34}\sigma_{44} \\ 2\sigma_{34}^2 & 2\sigma_{34}\sigma_{44} & 2\sigma_{44}^2 \end{pmatrix},$$

$$B_3 = \sigma_{11}\begin{pmatrix} \sigma_{33} & \sigma_{34} \\ \sigma_{34} & \sigma_{44} \end{pmatrix}, \qquad B_4 = \sigma_{22}\begin{pmatrix} \sigma_{33} & \sigma_{34} \\ \sigma_{34} & \sigma_{44} \end{pmatrix}.$$

Since Σ_0 is positive definite, all four blocks $B_1, \ldots, B_4$ are positive definite as well. This follows from exponential family theory but we can also calculate the leading principal minors that are all products of principal minors of Σ_0. Now compute block-diagonal matrices $B_i^{1/2}$ such that $B_i^{1/2}B_i^{T/2} = B_i$, for instance, by Cholesky decomposition. In particular, we can choose $B_1^{1/2}$ to be diagonal,

$$B_3^{1/2} = \sqrt{\sigma_{11}} \begin{pmatrix} \sigma_{33} & \sigma_{34} \\ \sigma_{34} & \sigma_{44} \end{pmatrix}^{1/2} \quad \text{and} \quad B_4^{1/2} = \sqrt{\sigma_{22}} \begin{pmatrix} \sigma_{33} & \sigma_{34} \\ \sigma_{34} & \sigma_{44} \end{pmatrix}^{1/2}.$$

Define $I(\Sigma_0)^{1/2}$ to be the block-diagonal matrix with blocks $B_i^{-1/2}$. This is a matrix square root satisfying $I(\Sigma_0)^{T/2}I(\Sigma_0)^{1/2} = I(\Sigma_0)$ as desired.

We claim that $I(\Sigma_0)^{1/2}$ leaves the tangent cone $TC_{\Sigma_0}(\Theta)$ invariant. Recall that in part 3 we showed that elements $\tau = (\tau_{ij})$ of $TC_{\Sigma_0}(\Theta)$ are of the form

$$\tau = \begin{pmatrix} \tau_{11} & 0 & \tau_{13} & \tau_{14} \\ & \tau_{22} & \tau_{23} & \tau_{24} \\ & & \tau_{33} & \tau_{34} \\ & & & \tau_{44} \end{pmatrix}$$

where the τ_{ij} are any reals satisfying $\tau_{13}\tau_{24} = \tau_{23}\tau_{14}$. Let $\bar{\tau} = (\bar{\tau}_{ij}) = I(\Sigma_0)^{1/2}\tau$. It is clear that $\bar{\tau}_{12} = 0$ such that $\bar{\tau} \in TC_{\Sigma_0}(\Theta)$ if $\bar{\tau}_{13}\bar{\tau}_{24} = \bar{\tau}_{14}\bar{\tau}_{23}$. The latter equation holds because the matrix

$$\begin{pmatrix} \tilde{\tau}_{13} & \tilde{\tau}_{23} \\ \tilde{\tau}_{14} & \tilde{\tau}_{24} \end{pmatrix} = \begin{pmatrix} \sigma_{33} & \sigma_{34} \\ \sigma_{34} & \sigma_{44} \end{pmatrix}^{-1/2} \begin{pmatrix} \tau_{13} & \tau_{23} \\ \tau_{14} & \tau_{24} \end{pmatrix} \begin{pmatrix} \sqrt{\sigma_{11}} & \\ & \sqrt{\sigma_{22}} \end{pmatrix}$$

has rank no more than 1. Therefore, $I(\Sigma_0)^{1/2}TC_{\Sigma_0}(\Theta) \subseteq TC_{\Sigma_0}(\Theta)$. Similarly, $I(\Sigma_0)^{-1/2}TC_{\Sigma_0}(\Theta) \subseteq TC_{\Sigma_0}(\Theta)$ such that $I(\Sigma_0)^{1/2}TC_{\Sigma_0}(\Theta) = TC_{\Sigma_0}(\Theta)$ as claimed.

Summarizing the results, we have

$$\lambda_n \xrightarrow{D} \min_{\tau \in TC_{\Sigma_0}(\Theta)} \|Z - \tau\|^2$$

with $Z \sim \mathcal{N}(0, I)$. Given any matrix $Z \in \mathbb{R}^{4\times4}$, the matrix $\tau \in TC_{\Sigma_0}(\Theta)$ minimizing the above norm must be equal to Z in the entries $\tau_{11}, \tau_{22}, \tau_{33}, \tau_{34}, \tau_{44}$. Also, $\tau_{12} = 0$. Thus,

$$\min_{\tau \in TC_{\Sigma_0}(\Theta)} \|Z - \tau\|^2 = Z_{12}^2 + \min_{\tau' \in C_1} \|Z_{12\times34} - \tau'\|^2,$$

where C_1 is the set of 2×2-matrices of rank at most 1. By classical linear algebra, $\min_{\tau' \in C_1} \|Z_{12\times34} - \tau'\|^2$ is equal to the smaller of the two eigenvalues of $W = Z_{12\times34}Z_{12\times34}^T$. The distribution of this latter random matrix is known as a Wishart distribution with 2 degrees of freedom and scale matrix equal to the identity matrix; we denote it by $\mathcal{W}_2(Id_2)$. Therefore,

$$\lambda_n \xrightarrow{D} V + \min\{\lambda_1(W), \lambda_2(W)\}$$

where $V \sim \chi_1^2$ and $W \sim \mathcal{W}_2(Id_2)$ are independent, and $\lambda_1(W)$ and $\lambda_1(W)$ are the two eigenvalues of W.

6.5   Fisher Information for Multivariate Normals

The team consisted of Alex Engström and Johannes Rauh.

Problem. Consider the model given by all centered multivariate normal distributions $\mathcal{N}(0, \Sigma)$ on $\mathbb{R}^m$ with (symmetric) positive definite covariance matrix Σ. Let

$$p_{\Sigma}(x) = \frac{1}{\sqrt{(2\pi)^m \det(\Sigma)}} \exp\left\{ -\frac{1}{2} x^T \Sigma^{-1} x \right\}, \quad x \in \mathbb{R}^m,$$

be the probability density of $\mathcal{N}(0, \Sigma)$. Viewing Σ as an element of $\mathbb{R}^{\binom{m+1}{2}}$, the Fisher-information $I(\Sigma)$ is defined to be the $\binom{m+1}{2} \times \binom{m+1}{2}$-matrix whose elements are the expected values

$$I(\Sigma)_{ij,kl} = \mathrm{E}\left[\left(\frac{\partial}{\partial \sigma_{ij}} \log p_{\Sigma}(X) \right) \left(\frac{\partial}{\partial \sigma_{kl}} \log p_{\Sigma}(X) \right) \right]$$

for $X \sim \mathcal{N}(0, \Sigma)$. Alternatively, one can compute the entries of the Fisher-information as

$$I(\Sigma)_{ij,kl} = -\mathrm{E}\left[\frac{\partial^2}{\partial \sigma_{ij} \partial \sigma_{kl}} \log p_{\Sigma}(X) \right]. \tag{6.5.1}$$

(This can be shown to be true quite generally, as long as the order of integration and differentiation can be interchanged.)

Verify that the inverse of the Fisher-information has entries of the form

$$\left(I(\Sigma)^{-1} \right)_{ij,kl} = \sigma_{ik}\sigma_{jl} + \sigma_{il}\sigma_{jk}. \tag{6.5.2}$$

Solution. We compute the Fisher-information matrix using formula (6.5.1), which decomposes as

$$I(\Sigma)_{ij,kl} = A_{ij,kl} + B_{ij,kl}, \tag{6.5.3}$$

where

$$A_{ij,kl} = \frac{1}{2} \frac{\partial^2}{\partial \sigma_{ij} \partial \sigma_{kl}} \log \det(\Sigma) \tag{6.5.4}$$

and

$$B_{ij,kl} = -\frac{1}{2} \mathrm{E}\left[\frac{\partial^2}{\partial \sigma_{ij} \partial \sigma_{kl}} X^T \Sigma^{-1} X \right]. \tag{6.5.5}$$

We denote the inverse of a matrix $\Sigma = (\sigma_{ij})$ by $\Sigma^{-1} = (\bar{\sigma}_{ij})$, and use the Einstein summation convention which states that one should sum over indices occurring twice in a product of variables. Further δ_{ij} denotes the Kronecker delta. For example, the fact that $\Sigma\Sigma^{-1}$ equals the identity matrix then corresponds to

$$\sigma_{kp}\bar{\sigma}_{pl} = \delta_{kl}. \tag{6.5.6}$$

The matrices Σ and Σ^{-1} are symmetric, and we identify the variables σ_{ij} and σ_{ji}, $i \neq j$. Differentiating (6.5.6) with respect to σ_{ij} we get

$$(\delta_{ik}\delta_{jp} + \delta_{ip}\delta_{jk})\bar{\sigma}_{pl} - \delta_{ij}\delta_{ik}\bar{\sigma}_{il} = -\sigma_{kp}\frac{\partial\bar{\sigma}_{pl}}{\partial\sigma_{ij}}.$$

The left-hand side comes from differentiating the first factor. The first two terms correspond to the two occurrences of σ_{ij} in Σ for $i \neq j$. If $i = j$ there is only one occurrence, so this is corrected by the last term on the left-hand side. By multiplying with $\bar{\sigma}_{qk}$ (and summing over k), we get the useful formula

$$\frac{\partial\bar{\sigma}_{ql}}{\partial\sigma_{ij}} = \delta_{ij}\bar{\sigma}_{qi}\bar{\sigma}_{il} - (\bar{\sigma}_{qi}\bar{\sigma}_{jl} + \bar{\sigma}_{il}\bar{\sigma}_{qj}).$$

In order to calculate $B_{ij,kl}$ in (6.5.5) we compute

$$\frac{\partial}{\partial\sigma_{ij}}\left(-\frac{X_q\bar{\sigma}_{ql}X_l}{2}\right) = \frac{1}{2}\left(X_q\bar{\sigma}_{qi}\bar{\sigma}_{jl}X_l + X_l\bar{\sigma}_{li}\bar{\sigma}_{jq}X_q - \delta_{ij}X_q\bar{\sigma}_{qi}\bar{\sigma}_{il}X_l\right)$$

$$= \left(1 - \frac{\delta_{ij}}{2}\right)X_p\bar{\sigma}_{pi}\bar{\sigma}_{jq}X_q.$$

Differentiating once again with respect to σ_{kl} we get

$$\left(1 - \frac{\delta_{ij}}{2}\right)\left(X_p[\delta_{kl}\bar{\sigma}_{pk}\bar{\sigma}_{li} - (\bar{\sigma}_{pk}\bar{\sigma}_{li} + \bar{\sigma}_{ik}\bar{\sigma}_{lp})]\bar{\sigma}_{jq}X_q\right.$$

$$\left. + X_q\bar{\sigma}_{qi}[\delta_{kl}\bar{\sigma}_{jk}\bar{\sigma}_{lp} - (\bar{\sigma}_{jk}\bar{\sigma}_{lp} + \bar{\sigma}_{pk}\bar{\sigma}_{lj})]X_p\right).$$

Taking expectations using that $E[X_pX_q] = \sigma_{pq}$ we find that

$$B_{ij,kl} = 2\left(1 - \frac{\delta_{ij}}{2}\right)(\delta_{kl}\bar{\sigma}_{jk}\bar{\sigma}_{li} - \bar{\sigma}_{li}\bar{\sigma}_{jk} - \bar{\sigma}_{ik}\bar{\sigma}_{jl}).$$

Now consider the term $A_{ij,kl}$ in (6.5.4). To calculate it we need to know the derivative of the determinant of a matrix A by its (i,j)-th element. By an expansion of the determinant, this equals the determinant of the submatrix $\hat{A}_{ij}$ obtained after removing the i-th row and j-th column with an appropriate sign. If $i \neq j$, then symmetry gives another term with i and j exchanged. Thus we get

$$\frac{\partial\log\det(\Sigma)}{\partial\sigma_{ij}} = (-1)^{i+j}\frac{\det(\hat{\Sigma}_{ij}) + \det(\hat{\Sigma}_{ji}) - \delta_{ij}\det(\hat{\Sigma}_{ij})}{\det(\Sigma)} = (2 - \delta_{ij})\bar{\sigma}_{ij}.$$

In the last step we have used the equality $\det(\hat{\Sigma}_{ij}) = \det(\hat{\Sigma}_{ji})$ and Cramer's rule for matrix inversion. Differentiating with respect to σ_{kl} we obtain that

$$A_{ij,kl} = \frac{1}{2}\frac{\partial^2}{\partial\sigma_{ij}\partial\sigma_{kl}}\log\det(\Sigma) = \frac{1}{2}\left(\bar{\sigma}_{ik}\bar{\sigma}_{jl} + \bar{\sigma}_{il}\bar{\sigma}_{jk} - \delta_{kl}\bar{\sigma}_{ik}\bar{\sigma}_{jl}\right)(2 - \delta_{ij}).$$

Note that $B_{ij,kl} = -2A_{ij,kl}$. Thus, according to (6.5.3),

$$I(\Sigma)_{ij,kl} = -A_{ij,kl} - B_{ij,kl} = A_{ij,kl}.$$

Now in order to verify the claim in (6.5.2), we need to show that the matrix with entries $A_{ij,kl}$ is indeed the inverse of the matrix M with entries

$$M_{ij,kl} = \sigma_{ik}\sigma_{jl} + \sigma_{il}\sigma_{jk}.$$

But

$$A_{ij,kl}M_{kl,pq} = (2 - \delta_{ij})\left(\delta_{ip}\delta_{jq} + \delta_{iq}\delta_{jp} - \sum_k \bar{\sigma}_{ik}\bar{\sigma}_{jk}\sigma_{kq}\sigma_{kp}\right) \qquad (6.5.7)$$

does not look like the result we want, and indeed one last ingredient is missing. Because of symmetry,

$$(I(\Sigma)M)_{ij,pq} = \sum_{k \leq l} A_{ij,kl}M_{kl,pq} \neq \sum_{k,l} A_{ij,kl}M_{kl,pq} = A_{ij,kl}M_{kl,pq}.$$

Instead,

$$\sum_{k \leq l} A_{ij,kl}M_{kl,pq} = \frac{1}{2}\left(\sum_{k,l} A_{ij,kl}M_{kl,pq} + \sum_{k=l} A_{ij,kl}M_{kl,pq}\right).$$

The correction term is

$$\sum_{k=l} A_{ij,kl}M_{kl,pq} = (2 - \delta_{ij})\sum_k \bar{\sigma}_{ik}\bar{\sigma}_{jk}\sigma_{kp}\sigma_{kq}. \qquad (6.5.8)$$

Putting (6.5.7) and (6.5.8) together, we obtain

$$(I(\Sigma)M)_{ij,pq} = (\delta_{iq}\delta_{jp} + \delta_{ip}\delta_{jq})\frac{2 - \delta_{ij}}{2}.$$

This might still look strange at first sight but it is actually what we are looking for. Simply check that the right-hand side is 1 if $\{i, j\} = \{p, q\}$ and zero otherwise.

6.6 The Intersection Axiom and Its Failure

The team consisted of Thomas Friedrich, Anna Kedzierska and Zhiqiang Xu.

Problem. Let $X_1 \in [r_1]$, $X_2 \in [r_2]$, and $X_3 \in [r_3]$ be discrete random variables. If all joint probabilities $p_{ijk} = P(X_1 = i, X_2 = j, X_3 = k)$ are positive, then the Intersection Axiom (Proposition 3.1.3) implies that

$$X_1 \perp\!\!\!\perp X_2 \mid X_3 \text{ and } X_1 \perp\!\!\!\perp X_3 \mid X_2 \implies X_1 \perp\!\!\!\perp (X_2, X_3). \qquad (6.6.1)$$

However, this implication might fail if some events are allowed to have zero probability. This problem explores the ways that the intersection axiom might fail, together with a characterization of zero patterns that still allow the conclusion of the intersection axiom to be deduced.

1. Show that the probability distribution with $p_{111} = p_{222} = \frac{1}{2}$, and $p_{ijk} = 0$ otherwise, satisfies $X_1 \perp\!\!\!\perp X_2 | X_3$ and $X_1 \perp\!\!\!\perp X_3 | X_2$ but not $X_1 \perp\!\!\!\perp (X_2, X_3)$.

2. Compute the primary decomposition of the conditional independence ideal $I_{1 \perp\!\!\!\perp 2|3} + I_{1 \perp\!\!\!\perp 3|2}$ for various choices of the number of states of the random variables.

3. Give a combinatorial description of the minimal primes of $I_{1 \perp\!\!\!\perp 2|3} + I_{1 \perp\!\!\!\perp 3|2}$. In particular, how many minimal primes are there?

4. Let (p_{ijk}) be a probability density satisfying $X_1 \perp\!\!\!\perp X_2 | X_3$ and $X_1 \perp\!\!\!\perp X_3 | X_2$. Let G be the bipartite graph with vertex set $[r_2] \cup [r_3]$, and an edge (j, k) if the marginal probability $p_{+jk} = \sum_i p_{ijk}$ is positive. Show that if G is connected then $X_1 \perp\!\!\!\perp (X_2, X_3)$. Conversely, if G is disconnected it need not be true that $X_1 \perp\!\!\!\perp (X_2, X_3)$.

Solution. 1. For a fixed value k of random variable X_3, form the $r_1 \times r_2$-matrix $M_k := (p_{ijk})_{i,j}$. This matrix has at most one non-zero entry and thus has rank less than or equal to 1. By Proposition 3.1.4, it holds that $X_1 \perp\!\!\!\perp X_2 | X_3$. By symmetry, we also have $X_1 \perp\!\!\!\perp X_3 | X_2$.

Let

$$P = \begin{pmatrix} p_{111} & p_{112} & \cdots & p_{11r_3} & \cdots & p_{1r_2r_3} \\ \vdots & \vdots & & \vdots & & \vdots \\ p_{r_111} & p_{r_112} & \cdots & p_{r_11r_3} & \cdots & p_{r_1r_2r_3} \end{pmatrix}. \tag{6.6.2}$$

This matrix has exactly two non-zero entries that appear in the invertible submatrix

$$\begin{pmatrix} p_{111} & p_{122} \\ p_{211} & p_{222} \end{pmatrix} = \begin{pmatrix} 1/2 & 0 \\ 0 & 1/2 \end{pmatrix}.$$

Hence, $\text{rank}(P) = 2$, which implies that X_1 is not independent of (X_2, X_3).

2. The following `Singular` code computes the primary decomposition of the conditional independence ideal $I_C = I_{1 \perp\!\!\!\perp 2|3} + I_{1 \perp\!\!\!\perp 3|2}$ for binary random variables.

```
LIB "primdec.lib";
ring R=0,(p(1..2)(1..2)(1..2)),dp;
matrix M1[2][2]=p(1..2)(1..2)(1);
matrix M2[2][2]=p(1..2)(1..2)(2);
matrix N1[2][2]=p(1..2)(1)(1..2);
matrix N2[2][2]=p(1..2)(2)(1..2);
ideal J=minor(M1,2),minor(M2,2);
ideal I=minor(N1,2),minor(N2,2);
primdecGTZ(I+J);
```

Running the code we find that there are three irreducible components. The following table summarizes our findings on the number of components when varying the number of states of the random variables:

(r_1, r_2, r_3)	(2,2,2)	(2,2,3)	(2,2,4)	(2,2,5)	(2,3,3)	(3,2,2)	(3,3,3)
#comp's	3	7	15	31	25	3	25

3. Based on the computations from part 2, Dustin Cartwright and Alexander Engström made the following conjecture. It involves complete bipartite graphs on a set of nodes $[p] \cup [q]$, which we denote by $K_{p,q}$.

Conjecture. *The minimal primes of $I_{1 \perp\!\!\!\perp 2|3} + I_{1 \perp\!\!\!\perp 3|2}$ correspond to the subgraphs of K_{r_2,r_3} that have the same vertex set $[r_2] \cup [r_3]$ and that have all connected components isomorphic to some complete bipartite graph $K_{p,q}$ with $p, q \geq 1$.*

Let $\eta(p, q)$ be the number of such subgraphs. Then the exponential generating function

$$f(x,y) \;=\; \sum_{p \geq 0, q \geq 0} \frac{\eta(p,q) x^p y^q}{p! q!}$$

is equal to

$$f(x,y) \;=\; e^{(e^x - 1)(e^y - 1)}.$$

This can be proved with Stirling numbers as described in [85].

4. Since $X_1 \perp\!\!\!\perp X_2 | X_3$ and $X_1 \perp\!\!\!\perp X_3 | X_2$, the matrices $M_k = (p_{ijk})_{i,j}$ and $N_j = (p_{ijk})_{i,k}$ have rank ≤ 1 for all $k \in [r_3]$ and $j \in [r_2]$, respectively. Suppose the graph G defined in the problem is connected. In order to prove that $X_1 \perp\!\!\!\perp (X_2, X_3)$, we need to show that each 2×2 minor of the matrix P in (6.6.2) is zero, or equivalently that any non-zero column in P is a multiple of any other non-zero column (recall Proposition 3.1.4). Since we are working with probability distributions, all entries of P are non-negative, and the non-zero columns correspond to the edges in G.

Consider two non-zero columns P_{jk} and $P_{j'k'}$ of P. Since G is connected, there is a sequence of edges $jk = j_1 k_1, j_2 k_2, \ldots, j_l k_l = j'k'$ such that each adjacent pair of edges share a vertex. At the first step, for instance, either $j_2 = j_1$ or $k_2 = k_1$. It follows that columns $P_{j_t k_t}$ and $P_{j_{t+1} k_{t+1}}$ both belong to one of the matrices M_k or N_j. Since both columns are non-zero, there is a constant $c_t \neq 0$ such that $P_{j_{t+1} k_{t+1}} = c_t P_{j_t k_t}$. Hence, $P_{j'k'} = P_{jk} \cdot \prod c_t$, which shows that P has rank 1.

If, on the other hand, the graph G is disconnected, then we can construct a counterexample similar to the one discussed in part 1. Let $C \subsetneq [r_2] \cup [r_3]$ be a connected component of G. Pick a table of joint probabilities such that the only non-zero entries are p_{1jk} for adjacent vertices $j, k \in C$ and p_{2jk} for adjacent vertices $j, k \notin C$. These joint probabilities are compatible with the graph G. Moreover, it holds that $X_1 \perp\!\!\!\perp X_2 | X_3$ because each matrix $M_k = (p_{ijk})_{i,j}$ has only one non-zero row, namely, the first row if $k \in C$ and the second row if $k \notin C$. Similarly, we have $X_1 \perp\!\!\!\perp X_3 | X_2$ because only the first row of N_j is non-zero if $j \in C$ and only the second row is non-zero if $j \notin C$. However, it does not hold that $X_1 \perp\!\!\!\perp (X_2, X_3)$. To see this, pick distinct indices $j, k \in C$ and $l, m \notin C$. Then

$$p_{1jk}p_{2lm} - p_{1jm}p_{2lk} = p_{1jk}p_{2lm} > 0$$

is a non-zero 2×2 minor of the matrix P in (6.6.2).

6.7 Primary Decomposition for CI Inference

The team consisted of Hugo Maruri-Aguilar, Filip Cools and Helene Neufeld.

Problem. In Examples 3.1.7 and 3.1.15 we saw that for either binary or normal random variables X_1, X_2, and X_3,

$$X_1 \perp\!\!\!\perp X_3 \text{ and } X_1 \perp\!\!\!\perp X_3 | X_2 \implies X_1 \perp\!\!\!\perp (X_2, X_3) \text{ or } (X_1, X_2) \perp\!\!\!\perp X_3. \qquad (6.7.1)$$

The goal of this problem is to characterize the failure of implication (6.7.1) when X_1, X_2, and X_3 are discrete.

1. Show that the implication (6.7.1) holds for discrete random variables as long as X_2 is binary.

2. Show that the implication (6.7.1) need not hold when X_2 is ternary (try computing a primary decomposition in `Singular`).

3. How many minimal primes does the ideal $I_{1 \perp\!\!\!\perp 3} + I_{1 \perp\!\!\!\perp 3 | 2}$ have for discrete random variables? (Note: the number of components will depend on the number of states of the random variables.)

Solution. 1. Let $P = (p_{ijk})_{i,j,k}$ be the probability distribution of discrete random variables $X_1 \in [r_1]$, $X_2 \in [r_2]$ and $X_3 \in [r_3]$. We consider the case where X_2 is binary, so $r_2 = 2$. In this case, we are going to prove (6.7.1).

Let $A_1, A_2 \in \mathbb{R}^{r_1 \times r_3}$ be the slices of P corresponding to X_2 equals 1 or 2, respectively. Since P satisfies both marginal and conditional independence, we have that A_1, A_2, and $A_1 + A_2$ have rank ≤ 1. So we can write $A_i = x_i y_i^T$ where $x_i \in \mathbb{R}^{r_1 \times 1}$ and $y_i \in \mathbb{R}^{r_3 \times 1}$ for $i = 1, 2$. Denote $N = (x_1, x_2) \in \mathbb{R}^{r_1 \times 2}$ and $M = (y_1, y_2) \in \mathbb{R}^{r_3 \times 2}$. We have that $NM^T = A_1 + A_2$ has rank 1.

We claim that N or M has rank ≤ 1. Indeed, assume N and M both have rank 2. So there exist $i_1, i_2 \in [r_1]$ and $j_1, j_2 \in [r_3]$ such that the submatrices $N(i_1, i_2) \in \mathbb{R}^{2 \times 2}$ of N, with row indices i_1, i_2 and $M(j_1, j_2) \in \mathbb{R}^{2 \times 2}$ of M, with row indices j_1, j_2 have non-zero determinant. The submatrix of $A_1 + A_2$ corresponding to the rows i_1, i_2 and columns j_1, j_2 is equal to

$$N(i_1, i_2)M(j_1, j_2)^T \in \mathbb{R}^{2 \times 2},$$

thus it has non-zero determinant. This is in contradiction with $\text{rank}(A_1 + A_2) = 1$.

If N has rank 1, the column vectors x_1 and x_2 are linearly dependent, so we have $x_2 = \alpha x_1$ for some $\alpha \in \mathbb{R}$. The matrix $(A_1, A_2) \in \mathbb{R}^{r_1 \times (2r_3)}$ can be written as

$$(A_1, A_2) = x_1(y_1^T, \alpha y_2^T),$$

hence it has rank equal to 1. This implies $X_1 \perp\!\!\!\perp (X_2, X_3)$. Analogously, we can show that $\mathrm{rank}(M) = 1$ implies $(X_1, X_2) \perp\!\!\!\perp X_3$.

2. The implication (6.7.1) does not always hold. For instance, consider the case where X_1 and X_3 are binary and X_2 is ternary. Using `Singular`, we can compute the primary decomposition of the CI ideal $I = I_{1 \perp\!\!\!\perp 3} + I_{1 \perp\!\!\!\perp 3|2}$.

```
LIB "primdec.lib";
ring r=0,(p(1..2)(1..3)(1..2)),dp;
matrix slZ1[2][2]=p(1..2)(1)(1..2);
matrix slZ2[2][2]=p(1..2)(2)(1..2);
matrix slZ3[2][2]=p(1..2)(3)(1..2);
matrix slZ[2][2]=slZ1+slZ2+slZ3;
ideal I=minor(slZ1,2),minor(slZ2,2),minor(slZ3,2),minor(slZ,2);
dim(std(I));
8
list L=primdecGTZ(I);
size(L);
1
```

So the algebraic variety $V(I) \subset \mathbb{R}^{12}$ is 8-dimensional and I has only one minimal prime. By examining the list of primary components L, it follows that I is a prime ideal. We can also compute the dimensions of the varieties $V(I_{1 \perp\!\!\!\perp \{2,3\}})$ and $V(I_{\{1,2\} \perp\!\!\!\perp 3})$. The `Singular` code below computes $\dim(V(I_{1 \perp\!\!\!\perp \{2,3\}}))$.

```
matrix slY1[2][3]=p(1..2)(1..3)(1);
matrix slY2[2][3]=p(1..2)(1..3)(2);
ideal IX=minor(concat(slY1,slY2),2);
dim(std(IX));
7
```

Hence, the variety $V(I)$ contains $V(I_{1 \perp\!\!\!\perp \{2,3\}})$ and $V(I_{\{1,2\} \perp\!\!\!\perp 3})$ as subvarieties of codimension 1. To show that the implication does not hold, it thus suffices to find one probability distribution $p \in V(I)$ that is not in $V(I_{1 \perp\!\!\!\perp \{2,3\}})$ or $V(I_{\{1,2\} \perp\!\!\!\perp 3})$. One such probability distribution is

$$\begin{pmatrix} p_{111} & p_{112} & p_{121} & p_{122} & p_{131} & p_{132} \\ p_{211} & p_{212} & p_{221} & p_{222} & p_{231} & p_{232} \end{pmatrix} = \frac{1}{12} \begin{pmatrix} 1 & 2 & 2 & 1 & 0 & 0 \\ 0 & 0 & 2 & 1 & 1 & 2 \end{pmatrix}.$$

This shows the implication (6.7.1) does not hold in this case.

3. The number of minimal primes is given by the formula

$$\max\{1, \min\{r_1, r_2, r_3, r_1 + r_3 - r_2\}\}.$$

We will now explain what these minimal primes are, and how they are derived. Our explanation is geometric and describes the varieties $V(Q)$ associated to each minimal prime Q, rather than ideal generators of the minimal primes.

Let $P = (p_{ijk})$ be the $r_1 \times r_2 \times r_3$ probability tensor and denote by A_j the slice of P corresponding to $X_2 = j$, for all $j = 1, \ldots, r_2$. Since $X_1 \perp\!\!\!\perp X_3$ and $X_1 \perp\!\!\!\perp X_3 | X_2$, this implies that $A_1, \ldots, A_{r_2}, A_1 + \cdots + A_{r_2}$ have rank ≤ 1. For all $j = 1, \ldots, r_2$, we can write A_j as $x_j y_j^T$ where $x_j \in \mathbb{R}^{r_1 \times 1}$ and $y_j \in \mathbb{R}^{r_3 \times 1}$. Hence, the rank one matrix $A_1 + \cdots + A_{r_2}$ is equal to NM^T, where

$$N = (x_1, \ldots, x_{r_2}) \in \mathbb{R}^{r_1 \times r_2} \text{ and } M = (y_1, \ldots, y_{r_2}) \in \mathbb{R}^{r_3 \times r_2}.$$

We have reduced the problem of studying the minimal primes of I to studying the irreducible components of the set of pairs $(N, M^T) \in \mathbb{C}^{r_1 \times r_2} \times \mathbb{C}^{r_2 \times r_3}$ such that $NM^T \in \mathbb{C}^{r_1 \times r_3}$ has rank ≤ 1. This variety is an example of a *quiver variety* and most of the results in this section can be proven using the combinatorial *wiring diagram* technique for studying quiver varieties. A good reference for this material is Section 17.1 of [70]. In particular, Example 17.11 contains an outline for finding the irreducible components in this case. We will provide a description of the irreducible components and some of the ideas of the necessary proofs, but refer the reader to [70] for the needed results on quiver varieties.

Let $\nu : \mathbb{R}^{r_2} \to \mathbb{R}^{r_1}$ be the linear map defined by N, and let $\mu : \mathbb{R}^{r_3} \to \mathbb{R}^{r_2}$ be the map corresponding to M^T. Denote by $\tilde{V}_{s,t}$ the set of all joint probability distributions P of X_1, X_2, X_3 such that $X_1 \perp\!\!\!\perp X_3$, $X_1 \perp\!\!\!\perp X_3 | X_2$, $\dim(\text{image}(\nu)) = s$ and $\dim(\text{image}(\mu)) = t$. For $P \in \tilde{V}_{s,t}$, we have

$$
\begin{aligned}
\dim(\text{image}(\mu) + \ker(\nu)) - (r_2 - s) &= t - \dim(\ker(\nu) \cap \text{image}(\mu)) \\
&= \dim(\text{image}(\nu \circ \mu)) = \text{rank}(NM^T) = 1.
\end{aligned}
$$

Let $V_{s,t}$ be the (Zariski) closure of $\tilde{V}_{s,t}$. These have the following properties:

- The set $V_{s,t}$ is an irreducible variety. Here is the basic idea: If $\tilde{V}_{s,t}$ is non-empty, then there is a pair of matrices N, M with rank $N = s$, rank $M = t$ and rank $NM^T = 1$. Then any other pair of matrices N', M' satisfying these three rank conditions can be obtained from N and M by a simultaneous change of coordinates on $\mathbb{C}^{r_1}$, $\mathbb{C}^{r_2}$, and $\mathbb{C}^{r_3}$. That is, there exist for $i = 1, 2, 3$, $G_i \in GL_{r_i}$ such that $N' = G_1 N G_2$, and $M' = G_3 M G_2^{-T}$. Thus, $V_{s,t}$ is an orbit closure and is irreducible.

- The set $V_{s,t}$ is non-empty if and only if

$$0 < s \leq r_2, \ 0 < t \leq r_2, \ s \leq r_1, \ t \leq r_3 \text{ and } r_2 - s \geq t - 1.$$

These conditions are necessary and sufficient to guarantee that there exists an $r_1 \times r_2$ matrix of rank s, and an $r_2 \times r_3$ matrix of rank t, whose product has rank 1. These inequality conditions are equivalent to $0 < s \leq \min\{r_1, r_2\}$ and $0 < t \leq \min\{r_3, r_2, r_2 - s + 1\}$.

- If $V_{s,t}, V_{s',t'}$ are non-empty, then $V_{s,t} \subset V_{s',t'}$ if and only if $s \leq s'$ and $t \leq t'$. This follows because all matrices of rank $< s$ are contained in the closure of the set of matrices of rank s.

- The set $V_{s,t}$ gives rise to an irreducible component of $V(I)$ if and only if $V_{s,t} \neq \emptyset$, but $V_{s',t} = \emptyset$ for any $s' > s$ and $V_{s,t'} = \emptyset$ for any $t' > t$. This follows because the $V_{s,t}$ are always contained in $V(I)$ and the irreducible components must be maximal and every point $p \in V(I)$ corresponds to a pair M, N satisfying some rank conditions.

Using the above properties, we can give an explicit description of the irreducible components of $V(I)$, depending on the values of r_1, r_2, r_3.

1. If $r_1 + r_3 - r_2 \leq 1$, the set $V_{s,t}$ is non-empty if and only if $0 < s \leq r_1$ and $0 < t \leq r_3$. Hence the only irreducible component of $V(I)$ corresponds to V_{r_1,r_3}.

2. If $r_1 \leq r_2, r_3 \leq r_2$ but $r_1 + r_3 - r_2 > 1$, the set $V_{s,t}$ is non-empty if and only if $0 < s \leq r_1$ and $0 < t \leq \min\{r_3, r_2 - s + 1\}$. The components of $V(I)$ are $V_{r_2-r_3+1,r_3}, V_{r_2-r_3+2,r_3-1}, \ldots, V_{r_1,r_2-r_1+1}$.

3. If $r_1 < r_2 \leq r_3$, the set $V_{s,t}$ is non-empty if and only if $0 < s \leq r_1$ and $0 < t \leq r_2 - s + 1$. The components of $V(I)$ are $V_{1,r_2}, V_{2,r_2-1}, \ldots, V_{r_1,r_2-r_1+1}$.

4. If $r_3 < r_2 \leq r_1$, the set $V_{s,t}$ is non-empty if and only if $0 < s \leq r_2$ and $0 < t \leq \min\{r_3, r_2 - s + 1\}$. The components of $V(I)$ are $V_{r_2-r_3+1,r_3}, \ldots, V_{r_2,1}$.

5. If $r_1 \geq r_2$ and $r_3 \geq r_2$, the set $V_{s,t}$ is non-empty if and only if $0 < s \leq r_2$ and $0 < t \leq r_2 - s + 1$. The components of $V(I)$ are $V_{1,r_2}, V_{2,r_2-1}, \ldots, V_{r_2,1}$.

Counting the numbers of irreducible components in each of the five cases, we see that the number of minimal primes of I is

$$\max\{1, \min\{r_1, r_2, r_3, r_1 + r_3 - r_2\}\}.$$

Note that this exercise shows that the discrete conditional independence model associated to the collection $\mathcal{C} = \{1 \perp\!\!\!\perp 3|2, 1 \perp\!\!\!\perp 3\}$ satisfies the unirationality conclusion of Question 3.1.9. This is because orbit closures, like $V_{s,t}$, are unirational varieties.

6.8   An Independence Model and Its Mixture

The team for this problem consisted of Dustin Cartwright, Mounir Nisse, Christof Söger and Piotr Zwiernik.

Problem. Let $\mathcal{M}$ denote the independence model for four binary random variables, where the first two variables are identically distributed, and the last two variables are identically distributed.

1. Express $\mathcal{M}$ as a log-linear model and find its Markov basis.

2. Compare the degree of $\mathcal{M}$ with the maximum likelihood degree of $\mathcal{M}$.

3. How does one compute the marginal likelihood integrals for the model $\mathcal{M}$?

4. Let $\mathrm{Mixt}^2(\mathcal{M})$ denote the mixture model which consists of all distributions that are convex combinations of two distributions in $\mathcal{M}$. Describe the geometry of this semi-algebraic set.

5. Pick a data set and run the expectation maximization (EM) algorithm for $\mathrm{Mixt}^2(\mathcal{M})$ on your data set.

6. Compare the degree of $\mathrm{Mixt}^2(\mathcal{M})$ with the maximum likelihood degree of $\mathrm{Mixt}^2(\mathcal{M})$.

7. How does one compute marginal likelihood integrals for the model $\mathrm{Mixt}^2(\mathcal{M})$?

Solution. 1. In the solution to this problem, we use some of the notational conventions from Section 5.2. We describe $\mathcal{M}$ for the binary random variables X_1, X_2, X_3 and X_4 where the first two variables are identically distributed, and the last two variables are identically distributed. If we set $p_{ijkl} = P(X_1 = i, X_2 = j, X_1 = k, X_2 = l)$ where $i, j, k, l \in \{0, 1\}$ then, since the four variables are independent, we have

$$p_{ijkl} = P(X_1 = i) \cdot P(X_2 = j) \cdot P(X_3 = k) \cdot P(X_4 = l).$$

The first two variables are identically distributed, so we have $P(X_1 = 0) = P(X_2 = 0) = s_0$ and $P(X_1 = 1) = P(X_2 = 1) = s_1$. The same holds for the two last variables $P(X_3 = 0) = P(X_4 = 0) = t_0$ and $P(X_3 = 1) = P(X_4 = 1) = t_1$. Thus, for example, $p_{0000} = s_0^2 t_0^2$ and $p_{1101} = s_1^2 t_0 t_1$.

We now write out the matrix A defining the log-linear model $\mathcal{M}$ as follows. The rows correspond to the four model parameters s_0, s_1, t_0, and t_1. The columns correspond to the probabilities p_{ijkl}, listed in lexicographic order. The resulting matrix A equals

$$
\begin{array}{cccccccccccccccc}
\scriptstyle 0000 & \scriptstyle 0001 & \scriptstyle 0010 & \scriptstyle 0011 & \scriptstyle 0101 & \scriptstyle 0101 & \scriptstyle 0110 & \scriptstyle 0111 & \scriptstyle 1000 & \scriptstyle 1001 & \scriptstyle 1010 & \scriptstyle 1011 & \scriptstyle 1101 & \scriptstyle 1101 & \scriptstyle 1110 & \scriptstyle 1111
\end{array}
$$

$$
\begin{pmatrix}
2 & 2 & 2 & 2 & 1 & 1 & 1 & 1 & 1 & 1 & 1 & 1 & 0 & 0 & 0 & 0 \\
0 & 0 & 0 & 0 & 1 & 1 & 1 & 1 & 1 & 1 & 1 & 1 & 2 & 2 & 2 & 2 \\
2 & 1 & 1 & 0 & 2 & 1 & 1 & 0 & 2 & 1 & 1 & 0 & 2 & 1 & 1 & 0 \\
0 & 1 & 1 & 2 & 0 & 1 & 1 & 2 & 0 & 1 & 1 & 2 & 0 & 1 & 1 & 2
\end{pmatrix}.
$$

To get the Markov basis of this model we use 4ti2. We enter the matrix A into an input file that we call m.mat, and then we issue the command markov m. This generates the Markov basis in the file m.mar, where the elements are the row vectors. The first four lines of output are

```
27 16
0  0  0  0  0  0  0  0  0  0  0  0  0  1 -1  0
0  0  0  0  0  0  0  0  0  0  0  0  1 -2  0  1
0  0  0  0  0  0  0  1  0  0  0 -1  0  0  0  0
```

This means the Markov basis consists of 27 elements. The first three vectors in the output represent the binomials $p_{1101} - p_{1110}$, $p_{1100}p_{1111} - p_{1101}^2$ and $p_{0111} - p_{1011}$. These are minimal generators of the toric ideal of A. In total, the Markov basis consists of seven linear binomials and twenty binomials of degree 2.

2. Next we use the Markov basis computed by 4ti2 to describe the model $\mathcal{M}$ in Singular. To do this, we use the command output --binomial m.mar in 4ti2, where m.mar is the name of the file containing the Markov basis. This writes the Markov basis as a list of binomials. This is not exactly in the format used by Singular, but we can use "search and replace" to prepare our Singular input:

```
ring R = 0, (x(1..16)), dp;
ideal II =
x(14)-x(15),
x(13)*x(16)-x(14)^2,
x(8)-x(12),
...
ideal I = std(II);
degree(I);
```

Here $x(1), x(2), \ldots$ stand for $p_{0000}, p_{0001}, \ldots$. This code generates the output:

```
// dimension (proj.)  = 2
// degree (proj.)   = 8
```

We see that the degree of the model $\mathcal{M}$ equals 8. On the other hand, the maximum likelihood degree of $\mathcal{M}$ is just 1 because, for an independence model, the maximum likelihood estimates are given by a rational function in the data (compare Example 2.1.2):

$$\hat{s}_1 = 1 - \hat{s}_0 = \frac{(Au)_2}{(Au)_1 + (Au)_2},$$

$$\hat{t}_1 = 1 - \hat{t}_0 = \frac{(Au)_3}{(Au)_3 + (Au)_4}.$$

3. To compute the marginal likelihood integral of $\mathcal{M}$ we use the formula in Lemma 5.2.3. Our independence model has two groups of distributions with two random variables in each group, so that $m = 2, s_1 = s_2 = 2$. The variables are binary, that is, $r_1 = r_2 = 1$. Given a data set u with $n = \sum u_i$, the marginal likelihood is

$$\frac{n!}{u_1! \cdots u_{16}!} \cdot \int_\Theta \theta^b d\theta = \frac{n!}{u_1! \cdots u_{16}!} \cdot \frac{b_0^{(1)}! b_1^{(1)}! b_0^{(2)}! b_1^{(2)}!}{(2n+1)!},$$

where $\Theta = \Delta_1 \times \Delta_1$ is the unit square, and

$$\begin{pmatrix} b_0^{(1)} \\ b_1^{(1)} \\ b_0^{(2)} \\ b_1^{(2)} \end{pmatrix} = Au.$$

We shall evaluate the marginal likelihood integral for a particular data set u. But first we make a simplification. We have several identical columns in A corresponding to events with identical probabilities. We can combine the occurrences of these events to get a reduced model matrix $\tilde{A}$ of format 4×9:

$$\tilde{A} = \begin{pmatrix} 2 & 2 & 2 & 1 & 1 & 1 & 0 & 0 & 0 \\ 0 & 0 & 0 & 1 & 1 & 1 & 2 & 2 & 2 \\ 2 & 1 & 0 & 2 & 1 & 0 & 2 & 1 & 0 \\ 0 & 1 & 2 & 0 & 1 & 2 & 0 & 1 & 2 \end{pmatrix}.$$

To compute the exact value of an integral, we pick the data

$$\tilde{u} = (2, 7, 17, 3, 11, 19, 5, 13, 23) \tag{6.8.1}$$

and use the `Maple` library written by Shaowei Lin, which can be found at the website `http://math.berkeley.edu/~shaowei/integrals.html`. The command we used is

```
ML([2,2],[1,1],[2,7,17,3,11,19,5,13,23],mixed=false);
```

This quickly returns

$$\frac{5769688351700026479359961612393318613178875486264098816000 0}{3490560108776711553962464223924346795223950427407566579430169945247360 3}.$$

This rational number is approximately equal to $1.652940552 \cdot 10^{-12}$.

4. We will compute equations for the mixture model using `Singular`. To do this, we use the code above to create the unmixed log-linear model in the ring `R`. Then we create a new ring `S`, which has twice as many parameters, corresponding to drawing independent observations from the model. We create three ring homomorphisms from `R` to `S`, which correspond to projection onto the sum of the sets of parameters (mixtures), projection onto the first set of probabilities, and projection onto the second set of probabilities. This method works for arbitrary secant varieties [90].

```
ring S = 0, (y(1..16), z(1..16)), dp;
map f = R, y(1)+z(1), y(2)+z(2), y(3)+z(3), y(4)+z(4), y(5)+z(5),
    y(6)+z(6), y(7)+z(7), y(8)+z(8), y(9)+z(9), y(10)+z(10),
    y(11)+z(11), y(12)+z(12), y(13)+z(13), y(14)+z(14),
    y(15)+z(15), y(16)+z(16);
map g = R, y(1..16);
map h = R, z(1..16);
ideal product = g(I) + h(I);
product = std(product);

setring R;
ideal mixture = preimage(S, f, product);
mixture = std(mixture);
degree(mixture);
```

which gives the output:

```
// dimension (proj.)  = 5
// degree (proj.)     = 10
```

In particular, notice that $\text{Mixt}^2(\mathcal{M})$ has the expected dimension $5 = 2 \cdot 2 + 1$ obtained by a simple parameter count: there are two parameters s_0 and t_0 in each of the two copies of $\mathcal{M}$ and one mixing parameter. We note that the generators of the ideal of $\text{Mixt}^2(\mathcal{M})$ are the 3×3-subdeterminants of

$$\begin{pmatrix} p_{0000} & p_{0001} & p_{0100} & p_{0101} \\ p_{0010} & p_{0011} & p_{0110} & p_{0111} \\ p_{1000} & p_{1001} & p_{1100} & p_{1101} \\ p_{1010} & p_{1011} & p_{1110} & p_{1111} \end{pmatrix}$$

together with the seven linear relations $p_{0010} - p_{0001}, p_{1110} - p_{1101}, p_{1000} - p_{0100}, p_{1011} - p_{0111}, p_{1010} - p_{0110}, p_{1010} - p_{0101}, p_{1010} - p_{1001}$. See [90, Example 5.2].

The model $\text{Mixt}^2(\mathcal{M})$ is a semi-algebraic set that is strictly smaller than the intersection of its Zariski closure, the secant variety, and the probability simplex Δ_{15}. For example, we will consider the point in probability space defined by $p_{0100} = p_{1000} = 1/2$, and all other probabilities 0. We can compute the fiber of the secant variety over this point as follows:

```
ideal pt = 2*x(2) - 1, 2*x(3)-1, x(1), x(4), x(5), x(6), x(7),
x(8),x(9),x(10),x(11),x(12),x(13),x(14),x(15),x(16);
setring S;
ideal fiber = product + f(pt);
fiber = std(fiber);
```

Since the ideal `fiber` is not the unit ideal, this point is in the secant variety. On the other hand, the ideal defining the fiber contains several equations of the form `y(16)+z(16)`. For this equation to hold with probabilistically valid non-negative real numbers, we must have $y_{16} = z_{16} = 0$. By adding these equations to the ideal:

```
ideal realfiber = fiber + y(4) + y(5) + y(1) + y(6) + y(7) +
y(8) + y(9) + y(10) + y(11) + y(12) + y(13) + y(14) + y(16) +
y(15);
realfiber = std(realfiber);
realfiber;
```

we get that `realfiber` is the unit ideal, so our point is not in the statistical model, despite being in its Zariski closure.

5. In order to find the maximum likelihood estimates for the mixture model, we use the same data vector u as before. We implement the EM algorithm in Matlab. The algorithm consists of two steps, as described on page 22 of the ASCB book [73]. The hidden model consists of two independent distributions from the model $\mathcal{M}$, together with the mixing parameter. Here is the complete Matlab code:

```
function th=oberem
th = [3/5,1/5,2/3,1/3,2/5];
th0 = [0,0,0,0,0];
tab=[2 3 5; 7 11 13; 17 19 23];
tol=1e-8;
while (max(abs(th-th0))>tol)
    th0=th;
    [U1,U2]=oberE(tab,th0); % E step
    th=oberM(U1,U2); % M step
end

function [U1,U2]=oberE(tab,th0)
s=th0(1); t=th0(2); u=th0(3); w=th0(4); p=th0(5);
P1=    p*[s^2,2*s*(1-s),(1-s)^2]'*[t^2,2*t*(1-t),(1-t)^2];
P2=(1-p)*[u^2,2*u*(1-u),(1-u)^2]'*[w^2,2*w*(1-w),(1-w)^2];
U1= (P1.*tab)./(P1+P2);
U2= tab-U1;

function th = oberM(U1,U2)
n1=sum(sum(U1));
n2=sum(sum(U2));
s=(2*(U1(1,1)+U1(1,2)+U1(1,3))+(U1(2,1)+U1(2,2)+U1(2,3)))/(2*n1);
t=(2*(U1(1,1)+U1(2,1)+U1(3,1))+(U1(1,2)+U1(2,2)+U1(3,2)))/(2*n1);
u=(2*(U2(1,1)+U2(1,2)+U2(1,3))+(U2(2,1)+U2(2,2)+U2(2,3)))/(2*n2);
w=(2*(U2(1,1)+U2(2,1)+U2(3,1))+(U2(1,2)+U2(2,2)+U2(3,2)))/(2*n2);
p=n1/(n1+n2);
th=[s,t,u,w,p];
```

The E-step is realized by the function **oberE** which computes the expected counts U_1 and U_2 for the model in each of the two hidden states. The subsequent M-step in **oberM** uses the tables U_1 and U_2 to compute estimates of the parameters $\theta = [s, t, v, w, \pi]$. The algorithm repeats both steps until the parameters converge. Convergence is checked by while(max(abs(th-th0))>tol) in the main loop.

For the initial values of parameters $\theta_0 = [3/5, 1/5, 2/3, 1/3, 2/5]$ we get

$$\hat{\theta} = [0.2895, 0.2137, 0.1942, 0.7973, 0.6379].$$

Starting with different values, for example $\theta_0 = [1/5, 4/5, 2/5, 1/5, 2/5]$ we get

$$\hat{\theta} = [0.1942, 0.7973, 0.2895, 0.2137, 0.3621].$$

These two vectors describe essentially the same solution under the symmetry of swapping the two copies of the model and changing π to $1 - \pi$.

6. As computed above, the degree of the mixture model is 10. We do not know the maximum likelihood degree. Unfortunately, our **Singular** implementation of

Algorithm 2.2.9 does not terminate for this model. Further work will be needed to determine the ML degree for models such as this one.

7. To compute the marginal likelihood integral of the mixture model $\text{Mixt}^2(\mathcal{M})$ we use Shaowei Lin's `Maple` libary mentioned above. The relevant background is explained in Section 3.2. The command in this case is

$$\texttt{ML([2,2],[1,1],[2,7,17,3,11,19,5,13,23]);}$$

This computation takes much longer (about two hours) than in the unmixed case described in part 3 of this exercise. Eventually, `Maple` returns the exact value which is a fraction with a 195 digits numerator and a 205 digits denominator. This rational number is approximately equal to $4.373970139 \cdot 10^{-9}$.

Chapter 7

Open Problems

On the final day of the Seminar, we organized a short open problem session to highlight some of the open problems that appeared during the lectures and to give students an opportunity to present problems. This final chapter lists problems that arose during the session. Their order loosely follows the order of Chapters 1-5.

7.1 Symmetry and Verification of Markov Bases

Studying the output of software for computing Markov bases, such as 4ti2, often reveals that a large set of moves contains only a few different combinatorial types; see for instance Exercise 6.1. Thomas Kahle posed the following problem.

Problem. *Give an algorithm that computes the combinatorial types of Markov moves without computing the Markov basis itself.*

Johannes Rauh proposed a related problem.

Problem. *Given a candidate for a Markov basis, give an algorithm that can verify that it is a Markov basis without computing one from scratch. In particular, is there a polynomial time algorithm in the bit complexity of the input that determines whether or not a set of integral vectors is a Markov basis of the lattice they span?*

7.2 Equivariant Inclusions in Binary Graph Models

Let $G = (V, E)$ be an undirected graph. The *binary graph model* $\mathcal{B}(G)$ is the hierarchical log-linear model for $\#V$ binary random variables associated to the simplicial complex

$$\{v : v \in V\} \cup \{\{v, w\} : (v, w) \in E\}$$

given by all nodes and edges in G. Note that this notion of a graph model is in general different from that of a graphical model, which is the hierarchical log-linear model associated to the complex of cliques in G; recall Proposition 3.3.3.

The *Markov width* $\mu(G)$ of a graph G is the maximal degree of a minimal generator of the toric ideal I_G associated with the binary graph model $\mathcal{B}(G)$. By Theorem 1.3.6, $2\mu(G)$ is a bound on the one-norm of any move in a minimal Markov basis for the model $\mathcal{B}(G)$. Markov widths were studied in [31]. The following problem was proposed by Alexander Engström.

Problem. *Let G, H be graphs and Γ a group acting transitively on both: $\Gamma \subseteq Aut(G)$, $\Gamma \subseteq Aut(H)$. If there exists an equivariant inclusion*

$$G \stackrel{\Gamma}{\hookrightarrow} H,$$

show that $\mu(G) \leq \mu(H)$ and "explain" the additional generators in I_H.

A few comments are in order to clarify this problem. Let V be the vertex set of G. The action of the group Γ on G is a function that maps every pair (g, v) of an element $g \in \Gamma$ and a vertex $v \in V$ to a new vertex $gv \in V$ in such a way that (i) $(gh)(v) = g(hv)$ for all $g, h \in \Gamma$ and $v \in V$, and (ii) the identity $e \in \Gamma$ satisfies $ev = v$ for all $v \in V$. The group Γ being a subset of the group of automorphisms $Aut(G)$ means that if vw is an edge of G, then so is $(gv)(gw)$ for all $g \in \Gamma$. When saying that Γ acts transitively on G, we mean that for any two vertices $v, w \in V$ there exists a group element g such that $gv = w$.

The notation $G \hookrightarrow H$ is from algebraic topology and denotes that G is included in H, or more formally, that there is an injective function from the vertex set of G to the vertex set of H such that, if vw is an edge of G, then $f(v)f(w)$ is an edge of H. The inclusion is equivariant, indicated by the notation $G \stackrel{\Gamma}{\hookrightarrow} H$, if $f(gv) = gf(v)$ for all $g \in \Gamma$ and $v \in V$.

The following is a simple example of the described situation. Let G be the four-cycle graph with vertex set $V = \{1, 2, 3, 4\}$ and edges 12, 23, 34, and 41. Let Γ be the group generated by the cyclic permutation (1234). This group acts transitively on G, and if we pick H to be the complete graph on V, then we have an equivariant inclusion $G \stackrel{\Gamma}{\hookrightarrow} H$.

In [31] it was conjectured that the removal of an edge from a graph does not increase its Markov width. A proof of that conjecture would solve the Markov width part of this open problem. Unfortunately, however, there are graphs which only differ by one edge while having toric ideals that do not look like each other at all.

7.3 Optimal Scaling in Iterative Proportional Scaling

In Section 2.1 we presented the 'iterative proportional scaling' (IPS) algorithm, which can be used to compute maximum likelihood estimates in log-linear models. In the setup of Algorithm 2.1.9, the model is specified via a non-negative integer matrix $A \in \mathbb{N}^{d \times r}$ whose columns all sum to the same value a. This integer a enters the IPS algorithm in Step 2, where a rescaling factor is raised to the power $1/a$. Seth Sullivant asked the following question about this exponent.

Question. *Can the exponent $1/a$ in Step 2 of the IPS algorithm be replaced with a larger number while still maintaining convergence? If yes, how does this affect the convergence rate of the algorithm? Does there exist a connection between convergence rates and the maximum likelihood degree of the model?*

Note that the choice of the exponent is allowed to depend on the data vector u.

7.4 Maximum Likelihood Degree of Gaussian Cycles

For a positive integer m, let C_m be the undirected cycle with vertex set $[m]$ and edges $(1,2),(2,3),\ldots,(m,m-1),(1,m)$. Consider the undirected Gaussian graphical model associated with the graph C_m. Recall from Section 3.2 that this is the family of multivariate normal distributions $\mathcal{N}(\mu,\Sigma)$ with covariance matrix in

$$\left\{\Sigma \in PD_m \ : \ (\Sigma^{-1})_{ij} = 0 \text{ for } ij \notin E(C_m)\right\}.$$

According to Theorem 2.1.14, maximum likelihood estimation for this model is a matrix completion problem. Based on computations for small m, Mathias Drton and Seth Sullivant conjectured a formula for the maximum likelihood degree.

Conjecture. *The maximum likelihood degree of the undirected Gaussian graphical model associated with C_m, the cycle of length m, is*

$$(m-3)2^{m-2}+1, \quad m \geq 3.$$

More generally, one might hope to give combinatorial formulae for the ML degree for Gaussian graphical models associated with arbitrary graphs.

7.5 Bumpiness of Likelihood Surface

The likelihood function of hidden variable and mixture models may have several local maxima; see for instance Example 5.2.5. This observation is the topic of the next problem by Bernd Sturmfels, which concerns the independence model $\mathcal{M}_{X \perp\!\!\!\perp Y}$ for two discrete random variables X and Y that both take values in $[r]$.

Problem. *Construct pairs (u,s) of a data table $u \in \mathbb{N}^{r \times r}$ and an integer $s \geq 2$ for which the likelihood function of the mixture model $\mathrm{Mixt}^s(\mathcal{M}_{X \perp\!\!\!\perp Y})$ has many local maxima in the probability simplex Δ_{r^2-1}. How does the number of local maxima behave when s is fixed and r goes to infinity? Even the case $s=2$ is of interest.*

7.6 Surjectively Positive Mixture Models

Let a_i denote the i-th column of a matrix $A \in \mathbb{N}^{d \times r}$ whose columns sum to a fixed value a. Let

$$\phi_A\colon \mathbb{R}^d \to \mathbb{R}^r : (\theta_1,\ldots,\theta_d) \mapsto (\theta^{a_1},\ldots,\theta^{a_r})$$

be the monomial map determined by A. Define $V_{A,\geq 0} = \overline{\phi_A(\mathbb{R}^d_{\geq 0})}$ to be the nonnegative part of the toric variety associated with A. Seth Sullivant proposed the

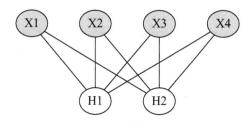

Figure 7.1: Undirected graph with two hidden and four observed nodes.

following problem about the relationship between mixture models and secant varieties. Recall Section 4.1 and in particular Example 4.1.2.

Problem. *Characterize pairs (A, s) with $A \in \mathbb{N}^{d \times r}$ and $s \in \mathbb{N} \setminus \{0, 1\}$ such that when taking the mixture of order s, it holds that*

$$\mathrm{Mixt}^s(V_{A, \geq 0}) = \mathrm{Sec}^s(V_{A, \geq 0}) \cap \mathbb{R}^r_{\geq 0}.$$

Furthermore, compute the tropicalization $\mathrm{Trop}(\mathrm{Mixt}^s(V_A))$ in these cases.

We remark that in [72], a rational map $g \colon \mathbb{R}^d \to \mathbb{R}^r$ such that

$$g(\mathbb{R}^d_{>0}) = f(\mathbb{R}^d) \cap \mathbb{R}^r_{>0}$$

is called *surjectively positive*. Thus, this problem asks for a characterization of the surjectively positive mixtures of toric varieties.

7.7 An Implicitization Challenge

Let $\mathcal{M} \subset \Delta_{15}$ be the hidden variable model associated with the undirected graph in Figure 7.1 when all variables are binary, X_1, X_2, X_3, X_4 are observed and H_1, H_2 are hidden. The model $\mathcal{M}$ can be parametrized by concatenating the monomial parametrization for the fully observed graphical model given in Proposition 3.3.3 and the marginalization map that takes the joint probabilities for $(X_1, X_2, X_3, X_4, H_1, H_2)$ to those for (X_1, X_2, X_3, X_4). In coordinates, we have

$$p_{i_1 i_2 i_3 i_4} = (a_{i_1} a_{i_2} a_{i_3} a_{i_4} + b_{i_1} b_{i_2} b_{i_3} b_{i_4}) \cdot (c_{i_1} c_{i_2} c_{i_3} c_{i_4} + d_{i_1} d_{i_2} d_{i_3} d_{i_4}).$$

Bernd Sturmfels proposed the following challenge.

Problem. *The model $\mathcal{M}$ is a hypersurface in Δ_{15}. Find its degree and defining polynomial.*

7.8 Finiteness in Factor Analysis

Consider the factor analysis model discussed in Section 4.2. Let $F_{m,s}$ be the covariance matrix parameter space and $I_{m,s}$ the associated ideal in the model with

m observed and s hidden variables. The following questions were raised by Mathias Drton, Bernd Sturmfels, and Seth Sullivant in [40]; see also the concluding paragraph of Section 4.2. For $a, b \in \mathbb{N}$ we abbreviate

$$\binom{[a]}{b} = \{A \subseteq [a] \ : \ \#A = b\}.$$

Question. *For each $s \geq 1$, does there exist $m_s \in \mathbb{N}$ such that for all $m \geq m_s$ it holds that*

$$\Sigma \in F_{m,s} \quad \Longleftrightarrow \quad \Sigma_{A \times A} \in F_{A,s} \text{ for all } A \in \binom{[m]}{m_s} \quad ?$$

A stronger ideal-theoretic version of this question would ask whether for each $s \geq 1$, there exists $m_s \in \mathbb{N}$ such that for all $m \geq m_s$ we have

$$I_{m,s} = \sum_{A \in \binom{[m]}{m_s}} I_{A,s}.$$

If $s = 1$, then it is not hard to show that $m_1 = 4$ is the smallest integer satisfying the set-theoretic property in question. The same holds for the ideal-theoretic property with $s = 1$; see [40, Theorem 16]. For general s, there is some evidence that the set-theoretic property might hold for $m_s = 2(s + 1)$ and larger; see [40, Prop. 30].

7.9 Rational Conditional Independence Models

As explained in Section 3.1, a conditional independence model for a random vector $X = (X_1, \ldots, X_m)$ is given by a set of conditional independence statements $\mathcal{C} = \{A_1 \perp\!\!\!\perp B_1 \mid C_1, A_2 \perp\!\!\!\perp B_2 \mid C_2, \ldots\}$. Here A_k, B_k, C_k are pairwise disjoint subsets of $[m]$ for each k. We also showed that conditional independence is an algebraic condition when X is discrete or multivariate normal; recall Propositions 3.1.4 and 3.1.13. Therefore, in either the Gaussian or the discrete case, the conditional independence statements $\mathcal{C}$ define an ideal

$$I_\mathcal{C} = I_{A_1 \perp\!\!\!\perp B_1 \mid C_1} + I_{A_2 \perp\!\!\!\perp B_2 \mid C_2} + \cdots.$$

The following question was posed by Seth Sullivant (compare Question 3.1.9). Recall that unirational varieties are varieties that have a rational parametrization.

Question. *Does every minimal prime (or irreducible component) of a Gaussian or discrete conditional independence ideal $I_\mathcal{C}$ correspond to a unirational variety?*

7.10 Discrete Chain Graph Models

Section 3.2 gave a brief introduction to graphical models based on chain graphs. These graphs may have both undirected and directed edges but no semi-directed cycles. In particular, we defined two Markov properties for chain graphs that we distinguished using the acronyms AMP and LWF. For discrete random variables, the AMP Markov property yields conditional independence models that are not yet well-understood. The below problems and questions, posed by Mathias Drton, concern their smoothness properties.

Consider the chain graph in Figure 3.2.5(a), which under the AMP Markov property, encodes the conditional independence statements $X_1 \perp\!\!\!\perp (X_2, X_4)$ and $X_2 \perp\!\!\!\perp X_4 \mid (X_1, X_3)$. If all variables are binary, then the conditional independence model given by these constraints is not smooth. Its singularities within the interior of the probability simplex Δ_{15} are exactly the points corresponding to complete independence of X_2, X_4 and the pair (X_1, X_3); compare Proposition 3.2.9.

Problem. *Describe the (positive part of the) singular locus of the discrete conditional independence model defined by $X_1 \perp\!\!\!\perp (X_2, X_4)$ and $X_2 \perp\!\!\!\perp X_4 \mid (X_1, X_3)$ when $X_1, \ldots, X_4$ have an arbitrary number of states. Does the positive part of the singular locus always include the locus of complete independence $X_2 \perp\!\!\!\perp X_4 \perp\!\!\!\perp (X_1, X_3)$?*

Over four variables, all chain graphs other than the one in Figure 3.2.5(a) determine discrete AMP chain graph models that are smooth over the interior of the probability simplex.

Problem. *Characterize the chain graphs for which all associated discrete AMP chain graph models are smooth over the interior of the probability simplex. Here 'all models' refers to allowing all combinations of integers ≥ 2 for the numbers of states of the considered random variables.*

7.11 Smoothness of Conditional Independence Models

In this book we encountered several examples of conditional independence models with singularities. For instance, in Example 3.1.7, Example 3.1.15 and Exercise 6.7, we saw that simultaneous marginal and conditional independence yields models that may break into several components and thus be singular where the components intersect. Another example occurred in Proposition 3.2.9, which concerned the conditional independence statements $X_1 \perp\!\!\!\perp (X_2, X_4)$ and $X_2 \perp\!\!\!\perp X_4 \mid (X_1, X_3)$. For binary variables, the associated conditional independence ideal is a prime ideal but the model is singular at positive distributions under which X_2, X_4 and the pair (X_1, X_3) are completely independent (compare also Problem 7.10).

In the examples just mentioned, the singular loci include distributions that exhibit additional conditional independence relations. In particular, the singularities of the above discrete models include the uniform distribution. This motivates the following question asked by Mathias Drton.

Question. *Suppose the uniform distribution defines a regular point in a discrete conditional independence model. Does this imply that the model is smooth over the interior of the probability simplex?*

The Gaussian analogue of this question is:

Question. *Suppose the standard multivariate normal distribution $\mathcal{N}(0, Id_m)$ defines a regular point in a Gaussian conditional independence model. Does it follow that the model is smooth over PD_m, the cone of positive definite matrices?*

7.12 Joint Entropies

Let $X \in [m]$ be a discrete random variable with distribution given by the probability vector $p \in \Delta_{m-1}$. The *entropy* of $p = (p_i \mid i \in [m])$ is

$$H(p) = -\sum_{i=1}^{m} p_i \log(p_i).$$

Let $X_1, \ldots, X_m$ be binary random variables. Their *joint entropy map* is the map

$$H : \Delta_{2^m - 1} \to \mathbb{R}_{\geq 0}^{2^m}$$
$$p \mapsto \left(H(p^A) \right)_{A \subseteq [m]},$$

where p^A denotes the probability vector for the distribution of $(X_i \mid i \in A)$. For example, if $m = 4$ and $p_{i+j+} = P(X_1 = i, X_3 = j)$, then

$$H(p^{\{1,3\}}) = -\sum_{i=1}^{2} \sum_{j=1}^{2} p_{i+j+} \log p_{i+j+}.$$

The following problem was posed by Bernd Sturmfels. See also [88, §4].

Problem. *Determine the image and the fibers of the joint entropy map $H : \Delta_{15} \to \mathbb{R}_{\geq 0}^{16}$ for four binary random variables.*

Bibliography

[1] A. Agresti, *Categorical Data Analysis*, Wiley Series in Probability and Mathematical Statistics: Applied Probability and Statistics, John Wiley & Sons Inc., New York, 1990.

[2] H. Akaike, *A new look at the statistical model identification*, IEEE Trans. Automatic Control **AC-19** (1974), 716–723.

[3] E. S. Allman and J. A. Rhodes, *Phylogenetic invariants for the general Markov model of sequence mutation*, Math. Biosci. **186** (2003), no. 2, 113–144.

[4] ———, *Molecular phylogenetics from an algebraic viewpoint*, Statist. Sinica **17** (2007), no. 4, 1299–1316.

[5] ———, *Phylogenetic ideals and varieties for the general Markov model*, Adv. in Appl. Math. **40** (2008), no. 2, 127–148.

[6] T. W. Anderson and H. Rubin, *Statistical inference in factor analysis*, Proceedings of the Third Berkeley Symposium on Mathematical Statistics and Probability, 1954–1955, vol. V, University of California Press, Berkeley and Los Angeles, 1956, pp. 111–150.

[7] S. A. Andersson, D. Madigan, and M. D. Perlman, *A characterization of Markov equivalence classes for acyclic digraphs*, Ann. Statist. **25** (1997), no. 2, 505–541.

[8] ———, *Alternative Markov properties for chain graphs*, Scand. J. Statist. **28** (2001), no. 1, 33–85.

[9] S. A. Andersson and M. D. Perlman, *Characterizing Markov equivalence classes for AMP chain graph models*, Ann. Statist. **34** (2006), no. 2, 939–972.

[10] S. Aoki and A. Takemura, *Minimal basis for a connected Markov chain over $3 \times 3 \times K$ contingency tables with fixed two-dimensional marginals*, Aust. N. Z. J. Stat. **45** (2003), no. 2, 229–249.

[11] M. Aoyagi and S. Watanabe, *Stochastic complexities of reduced rank regression in Bayesian estimation*, Neural Networks **18** (2005), no. 7, 924–933.

[12] V. I. Arnol'd, S. M. Guseĭn-Zade, and A. N. Varchenko, *Singularities of Differentiable Maps. Vol. II*, Monographs in Mathematics, vol. 83, Birkhäuser Boston Inc., Boston, MA, 1988.

[13] O. Barndorff-Nielsen, *Information and Exponential Families in Statistical Theory*, Wiley Series in Probability and Mathematical Statistics, John Wiley & Sons Ltd., Chichester, 1978.

[14] S. Basu, R. Pollack, and M.-F. Roy, *Algorithms in Real Algebraic Geometry*, second ed., Algorithms and Computation in Mathematics, vol. 10, Springer-Verlag, Berlin, 2006.

[15] N. Beerenwinkel and S. Sullivant, *Markov models for accumulating mutations*, arXiv:0709.2646, 2007.

[16] R. Benedetti and J.-J. Risler, *Real Algebraic and Semi-algebraic Sets*, Actualités Mathématiques. [Current Mathematical Topics], Hermann, Paris, 1990.

[17] P. Bickel and K. Doksum, *Mathematical Statistics. Vol 1*, Prentice-Hall, London, 2001.

[18] Y. M. M. Bishop, S. E. Fienberg, and P. W. Holland, *Discrete Multivariate Analysis: Theory and Practice*, The MIT Press, Cambridge, Mass.-London, 1975.

[19] L. D. Brown, *Fundamentals of Statistical Exponential Families with Applications in Statistical Decision Theory*, Institute of Mathematical Statistics Lecture Notes—Monograph Series, 9, Institute of Mathematical Statistics, Hayward, CA, 1986.

[20] H. Chernoff, *On the distribution of the likelihood ratio*, Ann. Math. Statistics **25** (1954), 573–578.

[21] R. Christensen, *Log-linear Models and Logistic Regression*, second ed., Springer Texts in Statistics, Springer-Verlag, New York, 1997.

[22] P. Clifford, *Markov random fields in statistics*, Disorder in physical systems, Oxford Sci. Publ., Oxford Univ. Press, New York, 1990, pp. 19–32.

[23] A. Conca, *Gröbner bases of ideals of minors of a symmetric matrix*, J. Algebra **166** (1994), no. 2, 406–421.

[24] R. G. Cowell, A. P. Dawid, S. L. Lauritzen, and D. J. Spiegelhalter, *Probabilistic Networks and Expert Systems*, Statistics for Engineering and Information Science, Springer-Verlag, New York, 1999.

[25] D. Cox, J. Little, and D. O'Shea, *Ideals, Varieties, and Algorithms*, third ed., Undergraduate Texts in Mathematics, Springer, New York, 2007.

[26] D. R. Cox and N. Wermuth, *Multivariate Dependencies*, Monographs on Statistics and Applied Probability, vol. 67, Chapman & Hall, London, 1996.

[27] J. N. Darroch and D. Ratcliff, *Generalized iterative scaling for log-linear models*, Ann. Math. Statist. **43** (1972), 1470–1480.

[28] L. J. Davis, *Exact tests for 2 × 2 contingency tables*, The American Statistician **40** (1986), no. 2, 139–141.

[29] J. A. De Loera and S. Onn, *Markov bases of three-way tables are arbitrarily complicated*, J. Symbolic Comput. **41** (2006), no. 2, 173–181.

[30] J. A. de Loera, B. Sturmfels, and R. R. Thomas, *Gröbner bases and triangulations of the second hypersimplex*, Combinatorica **15** (1995), no. 3, 409–424.

[31] M. Develin and S. Sullivant, *Markov bases of binary graph models*, Ann. Comb. **7** (2003), no. 4, 441–466.

[32] P. Diaconis and B. Efron, *Testing for independence in a two-way table: new interpretations of the chi-square statistic*, Ann. Statist. **13** (1985), no. 3, 845–913.

[33] P. Diaconis and B. Sturmfels, *Algebraic algorithms for sampling from conditional distributions*, Ann. Statist. **26** (1998), no. 1, 363–397.

[34] A. Dobra, *Markov bases for decomposable graphical models*, Bernoulli **9** (2003), no. 6, 1–16.

[35] A. Dobra and S. Sullivant, *A divide-and-conquer algorithm for generating Markov bases of multi-way tables*, Comput. Statist. **19** (2004), no. 3, 347–366.

[36] J. Draisma and J. Kuttler, *On the ideals of equivariant tree models*, arXiv:0712.3230, 2008.

[37] M. Drton, *Likelihood ratio tests and singularities*, arXiv:math/0703360, 2007.

[38] ———, *Discrete chain graph models*, Manuscript, 2008.

[39] M. Drton, H. Massam, and I. Olkin, *Moments of minors of Wishart matrices*, Annals of Statistics **36** (2008), no. 5, 2261–2283, arXiv:math/0604488.

[40] M. Drton, B. Sturmfels, and S. Sullivant, *Algebraic factor analysis: tetrads, pentads and beyond*, Probab. Theory Related Fields **138** (2007), no. 3-4, 463–493.

[41] M. Drton and S. Sullivant, *Algebraic statistical models*, Statist. Sinica **17** (2007), no. 4, 1273–1297.

[42] D. Edwards, *Introduction to Graphical Modelling*, second ed., Springer Texts in Statistics, Springer-Verlag, New York, 2000.

[43] D. Eisenbud, *Commutative Algebra. With a View Toward Algebraic Geometry*, Graduate Texts in Mathematics, vol. 150, Springer-Verlag, New York, 1995.

[44] D. Eisenbud and B. Sturmfels, *Binomial ideals*, Duke Math. J. **84** (1996), no. 1, 1–45.

[45] M. J. Evans, Z. Gilula, and I. Guttman, *Latent class analysis of two-way contingency tables by Bayesian methods*, Biometrika **76** (1989), no. 3, 557–563.

[46] J. Felsenstein, *Inferring Phylogenies*, Sinauer Associates, Inc., Sunderland, 2003.

[47] S. E. Fienberg, *Expanding the statistical toolkit with algebraic statistics (editorial)*, Statist. Sinica **17** (2007), no. 4, 1261–1272.

[48] H. Flenner, L. O'Carroll, and W. Vogel, *Joins and Intersections*, Springer Monographs in Mathematics, Springer-Verlag, Berlin, 1999.

[49] M. Frydenberg, *The chain graph Markov property*, Scand. J. Statist. **17** (1990), no. 4, 333–353.

[50] L. D. Garcia, M. Stillman, and B. Sturmfels, *Algebraic geometry of Bayesian networks*, J. Symbolic Comput. **39** (2005), no. 3-4, 331–355.

[51] D. Geiger, C. Meek, and B. Sturmfels, *On the toric algebra of graphical models*, Ann. Statist. **34** (2006), no. 3, 1463–1492.

[52] B. Georgi and A. Schliep, *Context-specific independence mixture modeling for positional weight matrices*, Bioinformatics **22** (2006), no. 14, e166–e173.

[53] H. Hara, A. Takemura, and R. Yoshida, *Markov bases for two-way subtable sum problems*, arXiv:0708.2312, 2007.

[54] ———, *A Markov basis for conditional test of common diagonal effect in quasi-independence model for two-way contingency tables*, arXiv:0802.2603, 2008.

[55] H. H. Harman, *Modern Factor Analysis*, revised ed., University of Chicago Press, Chicago, Ill., 1976.

[56] D. M. A. Haughton, *On the choice of a model to fit data from an exponential family*, Ann. Statist. **16** (1988), no. 1, 342–355.

[57] R. Hemmecke and R. Hemmecke, *4ti2 - Software for computation of Hilbert bases, Graver bases, toric Gröbner bases, and more*, 2003, available at http://www.4ti2.de.

[58] R. Hemmecke and K. Nairn, *On the Gröbner complexity of matrices*, arXiv:0708.4392, 2007.

[59] S. Hoşten and S. Sullivant, *Gröbner bases and polyhedral geometry of reducible and cyclic models*, J. Combin. Theory Ser. A **100** (2002), 277–301.

[60] S. Højsgaard and S. L. Lauritzen, *Graphical Gaussian models with edge and vertex symmetries*, J. Roy. Statist. Soc. Ser. B **70** (2008), no. 5, 1005–1027.

[61] S. Hoşten, A. Khetan, and B. Sturmfels, *Solving the likelihood equations*, Found. Comput. Math. **5** (2005), no. 4, 389–407.

[62] S. Hoşten and S. Sullivant, *A finiteness theorem for Markov bases of hierarchical models*, J. Combin. Theory Ser. A **114** (2007), no. 2, 311–321.

[63] A. N. Jensen, *Gfan, a software system for Gröbner fans*, available at http://www.math.tu-berlin.de/~jensen/software/gfan/gfan.html.

[64] T. L. Kelley, *Essential Traits of Mental Life*, Harvard University Press, Cambridge, MA, 1935.

[65] J. M. Landsberg and L. Manivel, *On the ideals of secant varieties of Segre varieties*, Found. Comput. Math. **4** (2004), no. 4, 397–422.

[66] J. M. Landsberg and J. Weyman, *On the ideals and singularities of secant varieties of Segre varieties*, Bull. Lond. Math. Soc. **39** (2007), no. 4, 685–697.

[67] S. Lauritzen, *Graphical Models*, Oxford University Press, New York, 1996.

[68] S. L. Lauritzen and N. Wermuth, *Graphical models for associations between variables, some of which are qualitative and some quantitative*, Ann. Statist. **17** (1989), no. 1, 31–57.

[69] S. Lin, B. Sturmfels, and Z. Xu, *Marginal likelihood integrals for mixtures of independence models*, arXiv:0805.3602, 2008.

[70] E. Miller and B. Sturmfels, *Combinatorial Commutative Algebra*, Graduate Texts in Mathematics, vol. 227, Springer-Verlag, New York, 2004.

[71] D. Mond, J. Smith, and D. van Straaten, *Stochastic factorizations, sandwiched simplices and the topology of the space of explanations*, R. Soc. Lond. Proc. Ser. A Math. Phys. Eng. Sci. **459** (2003), no. 2039, 2821–2845.

[72] L. Pachter and B. Sturmfels, *The tropical geometry of statistical models*, Proc. Natl. Acad. Sci. USA **101** (2004), 16132–16137.

[73] L. Pachter and B. Sturmfels (eds.), *Algebraic Statistics for Computational Biology*, Cambridge University Press, New York, 2005.

[74] G. Pistone, E. Riccomagno, and H. P. Wynn, *Algebraic Statistics*, Monographs on Statistics and Applied Probability, vol. 89, Chapman & Hall/CRC, Boca Raton, FL, 2001.

[75] R Development Core Team, *R: A language and environment for statistical computing*, R Foundation for Statistical Computing, Vienna, Austria, 2008.

[76] F. Rapallo, *Algebraic Markov bases and MCMC for two-way contingency tables*, Scand. J. Statist. **30** (2003), no. 2, 385–397.

[77] T. S. Richardson and P. Spirtes, *Ancestral graph Markov models*, Ann. Statist. **30** (2002), no. 4, 962–1030.

[78] C. P. Robert and G. Casella, *Monte Carlo Statistical Methods*, Springer Texts in Statistics, Springer-Verlag, New York, 1999.

[79] A. Roverato and M. Studený, *A graphical representation of equivalence classes of AMP chain graphs*, J. Mach. Learn. Res. **7** (2006), 1045–1078.

[80] D. Rusakov and D. Geiger, *Asymptotic model selection for naive Bayesian networks*, J. Mach. Learn. Res. **6** (2005), 1–35.

[81] F. Santos and B. Sturmfels, *Higher Lawrence configurations*, J. Combin. Theory Ser. A **103** (2003), no. 1, 151–164.

[82] G. Schwarz, *Estimating the dimension of a model*, Ann. Statist. **6** (1978), no. 2, 461–464.

[83] C. Semple and M. Steel, *Phylogenetics*, Oxford University Press, Oxford, 2003.

[84] J. Sidman and S. Sullivant, *Prolongations and computational algebra*, arXiv:math/0611696, 2006.

[85] R. P. Stanley, *Enumerative Combinatorics. Vol. 1*, Cambridge Studies in Advanced Mathematics, vol. 49, Cambridge University Press, Cambridge, 1997, Corrected reprint of the 1986 original.

[86] M. Studený, *Probabilistic Conditional Independence Structures*, Information Science and Statistics, Springer-Verlag, New York, 2005.

[87] B. Sturmfels, *Gröbner Bases and Convex Polytopes*, American Mathematical Society, Providence, 1995.

[88] B. Sturmfels, *Open problems in algebraic statistics*, Emerging Applications of Algebraic Geometry (M. Putinar and S. Sullivant, eds.), I.M.A. Volumes in Mathematics and its Applications, vol. 149, Springer, New York, 2008, pp. 351–364.

[89] B. Sturmfels and S. Sullivant, *Toric ideals of phylogenetic invariants*, Journal of Computational Biology **12** (2005), 204–228.

[90] B. Sturmfels and S. Sullivant, *Combinatorial secant varieties*, Pure Appl. Math. Q. **2** (2006), no. 3, 867–891.

[91] S. Sullivant, *Algebraic geometry of Gaussian Bayesian networks*, Adv. in Appl. Math. **40** (2008), no. 4, 482–513.

[92] _____, *A Gröbner basis for the secant ideal of the second hypersimplex*, arXiv:0804.2897, 2008.

[93] A. Takken, *Monte Carlo goodness-of-fit tests for discrete data*, Ph.D. thesis, Stanford University, 1999.

[94] A. W. van der Vaart, *Asymptotic Statistics*, Cambridge Series in Statistical and Probabilistic Mathematics, vol. 3, Cambridge University Press, Cambridge, 1998.

[95] S. Vavasis, *On the complexity of nonnegative matrix factorization*, arXiv:0708.4149, 2007.

[96] S. Watanabe, *Algebraic analysis for nonidentifiable learning machines*, Neural Computation **13** (2001), 899–933.

[97] _____, *Algebraic Geometry and Statistical Learning Theory*, Cambridge University Press, Cambridge, 2008.

[98] N. Wermuth and D. R. Cox, *Joint response graphs and separation induced by triangular systems*, J. R. Stat. Soc. Ser. B Stat. Methodol. **66** (2004), no. 3, 687–717.

[99] J. Whittaker, *Graphical Models in Applied Multivariate Statistics*, Wiley Series in Probability and Mathematical Statistics: Probability and Mathematical Statistics, John Wiley & Sons Ltd., Chichester, 1990.

[100] R. Wong, *Asymptotic Approximations of Integrals*, Academic Press, Boston, MA, 1989.

[101] K. Yamazaki and S. Watanabe, *Singularities in mixture models and upper bounds of stochastic complexity*, Neural Networks **16** (2003), 1029–1038.

[102] _____, *Newton diagram and stochastic complexity in mixture of binomial distributions*, Algorithmic Learning Theory, Lecture Notes in Comput. Sci., vol. 3244, Springer, Berlin, 2004, pp. 350–364.

[103] _____, *Algebraic geometry and stochastic complexity of hidden Markov models*, Neurocomputing **69** (2005), 62–84.

[104] F. L. Zak, *Tangents and Secants of Algebraic Varieties*, Translations of Mathematical Monographs, vol. 127, American Mathematical Society, Providence, RI, 1993.

[105] G. M. Ziegler, *Lectures on Polytopes*, Graduate Texts in Mathematics, vol. 152, Springer-Verlag, New York, 1995.

Oberwolfach Seminars (OWS)

The workshops organized by the *Mathematisches Forschungsinstitut Oberwolfach* are intended to introduce students and young mathematicians to current fields of research. By means of these well-organized seminars, also scientists from other fields will be introduced to new mathematical ideas. The publication of these workshops in the series *Oberwolfach Seminars* (formerly *DMV seminar*) makes the material available to an even larger audience.

OWS 39: Drton, M. / Sturmfels, B. / Sullivant, S., Lectures on Algebraic Statistics (2008). ISBN 978-3-7643-8904-8

OWS 38: Bobenko, A.I. / Schröder, P. / Sullivan, J.M. / Ziegler, G.M. (Eds.), Discrete Differential Geometry (2008). ISBN 978-3-7643-8620-7

Discrete differential geometry is an active mathematical terrain where differential geometry and discrete geometry meet and interact. It provides discrete equivalents of the geometric notions and methods of differential geometry, such as notions of curvature and integrability for polyhedral surfaces. Current progress in this field is to a large extent stimulated by its relevance for computer graphics and mathematical physics. This collection of essays, which documents the main lectures of the 2004 Oberwolfach Seminar on the topic, as well as a number of additional contributions by key participants, gives a lively, multi-facetted introduction to this emerging field.

OWS 37: Galdi, G.P. / Rannacher, R. / Robertson, A.M. / Turek, S., Hemodynamical Flows (2008). ISBN 978-3-7643-7805-9

This book surveys results on the physical and mathematical modeling as well as the numerical simulation of hemodynamical flows, i.e., of fluid and structural mechanical processes occurring in the human blood circuit. The topics treated are continuum mechanical description, choice of suitable liquid and wall models, mathematical analysis of coupled models, numerical methods for flow simulation, parameter identification and model calibration, fluid-solid interaction, mathematical analysis of piping systems, particle transport in channels and pipes, artificial boundary conditions, and many more. Hemodynamics is an area of active current research, and this book provides an entry into the field for graduate students and researchers.

OWS 36: Cuntz, J. / Meyer, R. / Rosenberg, J.M., Topological and Bivariant K-theory (2007). ISBN 978-3-7643-8398-5

Topological K-theory is one of the most important invariants for noncommutative algebras. Bott periodicity, homotopy invariance, and various long exact sequences distinguish it from algebraic K-theory. We describe a bivariant K-theory for bornological algebras, which provides a vast generalization of topological K-theory. In addition, we discuss other approaches to bivariant K-theories for operator algebras. As applications, we study K-theory of crossed products, the Baum-Connes assembly map, twisted K-theory with some of its applications, and some variants of the Atiyah-Singer Index Theorem.

OWS 35: Itenberg, I. / Mikhalkin, G. / Shustin, E., Tropical Algebraic Geometry (2007). ISBN 978-3-7643-8309-1

Tropical geometry is algebraic geometry over the semifield of tropical numbers, i.e., the real numbers and negative infinity enhanced with the $(max,+)$-arithmetics. Geometrically, tropical varieties are much simpler than their classical counterparts. Yet they carry information about complex and real varieties.
These notes present an introduction to tropical geometry and contain some applications of this rapidly developing and attractive subject. It consists of three chapters which complete each other and give a possibility for non-specialists to make the first steps in the subject which is not yet well represented in the literature. The intended audience is graduate, post-graduate, and Ph.D. students as well as established researchers in mathematics.

OWS 34: Lieb, E.H. / Seiringer, R. / Solovej, J.P. / Yngvason, J., The Mathematics of the Bose Gas and its Condensation (2005). ISBN 978-3-7643-7336-8

OWS 33: Kreck, M. / Lück, W., The Novikov Conjecture: Geometry and Algebra (2004). ISBN 978-3-7643-7141-8